AF334447

# LACAME 2006

# LACAME 2006

*Proceedings of the 10th Latin American Conference on the Applications of the Mössbauer Effect (LACAME 2006) held in Rio de Janeiro, Brazil, 5–9 November 2006*

*Edited by*

## C. LARICA

*UFES, Brazil*

## R. C. MERCADER

*UNLP, Argentina*

## C. PARTITI

*USP, Brazil*

and

## J. R. GANCEDO

*IQFR, Spain*

Reprinted from *Hyperfine Interactions*
Volume 175, Nos. 1–3 (2007)

**A C.I.P. Catalogue record for this book is available from the Library of Congress.**

ISBN 978-3-540-85558-3

Published by Springer
P.O. Box 990, 3300 AZ Dordrecht, The Netherlands

Sold and distributed in North, Central and South America
by Springer
101 Philip Drive, Norwell, MA 02061, U.S.A.

In all other countries, sold and distributed
by Springer
P.O. Box 990, 3300 AZ Dordrecht, The Netherlands

*Printed on acid-free paper*



Printed in The Netherlands

# Table of Contents

# Tenth Latin American Conference on the Applications of the Mössbauer Effect, LACAME 2006 Rio de Janeiro, Brazil, 5–9 November 2006

## Organizing Committee

E. Baggio Saitovitch (Chair)

*Centro Brasileiro de Pesquisas Físicas (CBPF)*
*Rio de Janeiro - Brazil*

E. C. Passamani (Vice-Chair)

*Universidade Federal do Espírito Santo (UFES)*
*Vitória - Brazil*

## Latin America Committee

| | |
|---|---|
| F. González-Jiménez | *Venezuela* |
| G. A. Pérez | *Colombia* |
| J. A. Jaén | *Panama* |
| N. R. Furet Bridón | *Cuba* |
| N. Nava | *Mexico* |
| R. C. Mercader | *Argentina* |
| V. A. Peña Rodriguez | *Peru* |

## National Committee

| | |
|---|---|
| Andréa Paesano | *UEM* |
| Carmem Partiti | *USP* |
| Diana Guenzburger | *CBPF* |
| José Domingo Fabris | *UFMG* |
| Rosa Bernstein Scorzelli | *CBPF* |
| Valderes Drago | *UFSC* |

## Local Committee

| | |
|---|---|
| Ada Lopez | *CBPF* |
| Dalber Sanchez Candela | *CBPF* |
| Izabel de Azevedo | *CBPF* |
| Jose Rafael Cápua Proveti | *UFES* |
| Magda Bittencourt | *CBPF* |

## SPONSORS

Centro Brasileiro de Pesquisas Físicas, CBPF, Brazil
Universidade Federal do Espírito Santo, UFES, Brazil
Fundação de Apoio a Pesquisa do Rio de Janeiro, FAPERJ, Brazil
Centro Latinoamericano de Física, CLAF, Rio de Janeiro, Brazil
Sociedade Brasileira de Física, SBF, São Paulo, Brazil
Coordenação de Aperfeiçoamento de Pessoal de Nível Superior, CAPES, Brazil
Abdus Salam International Centre for Theoretical Physics, ICTP, Trieste, Italy

**Preface**

The tenth Latin American Conference on the Application of the Mössbauer Effect (LACAME 2006) was held in the Rio de Janeiro city, Brazil, during November 5th–9th, 2006. The LACAME's have shown to be a very effective mechanism to improve scientific collaboration and dissemination of academic exchange among researchers and students from so many Latin American countries. Special effort has been done by the LACAME 2006 organizing committees in order to bring students and young researchers providing partial support for accommodation, transportation and registration fees. The LACAME 2006 scientific program have consisted of 17 invited talks, 21 oral presentations and 32 contributed paper in the form of posters. 26 manuscripts have been submitted for publication in the Conference Proceedings. The conference was attended by 112 registered participants from 14 countries, being 99 of them from Latin American countries.

All the presented contributions were related to the principal applications of the Mössbauer Effect such as: catalysis; applications in mineralogy, geology and soils; amorphous, nanocrystals and small particles; chemical applications: electronic structures, bounds; corrosion and environments; experimental techniques and data analysis; magnetism and magnetic materials; physical metallurgy and materials science.

Considering the large number of participants and the high quality of the oral and poster presentations the meeting has reached the main aims initially expected.

*C. Larica, UFRES, Brazil*
*R. C. Mercader, UNLP, Argentina*
*C. Partiti, USP, Brazil*
*J. R. Gancedo, IQFR, Spain*

Hyperfine Interact (2007) 175:1–8
DOI 10.1007/s10751-008-9580-5

# Improving detector signal processing with pulse height analysis in Mössbauer spectrometers

Jiri Pechousek · Miroslav Mashlan · Jiri Frydrych ·
Dalibor Jancik · Roman Prochazka

Published online: 20 March 2008


**Abstract** A plenty of different programming techniques and instrument solutions are used in the development of Mössbauer spectrometers. Each of them should provide a faster spectrum accumulation process, increased productivity of measurements, decreased nonlinearity of the velocity scale, etc. The well known virtual instrumentation programming method has been used to design a computer-based Mössbauer spectrometer. Hardware solution was based on two commercially-available PCI modules produced by National Instruments Co. Virtual Mössbauer spectrometer is implemented by the graphical programming language LabVIEW 7 Express. This design environment allows to emulate the multichannel analyzer on the digital oscilloscope platform. This is a novel method based on Waveform Peak Detection function which allows detailed analysis of the acquired signal. The optimal treatment of the detector signal from various detector types is achieved by mathematical processing only. As a result, the possibility of an increase of signal/noise ratio is presented.

**Keywords** Mössbauer spectrometers · Waveform Peak Detection · Digital signal processing

## 1 Introduction

At present, various programming languages have been used for the software implementation of Mössbauer spectrometers. A few years ago, the first application of current well-known virtual instrumentation programming method in Mössbauer spectrometry was published [1–3]. The advanced use of the graphical programming language LabVIEW [4] in Mössbauer spectrometry has been described in [5], where the hardware solution is based on two commercially-available PXI or PCI modules produced by National Instruments Co [6]. Data acquisition is realized via NI 5102 digital oscilloscope that is used as a multichannel analyzer, and NI 5401 function generator which is used as a velocity

J. Pechousek (✉) · M. Mashlan · J. Frydrych · D. Jancik · R. Prochazka
Department of Experimental Physics, Palacky University, Svobody 26, 771 46 Olomouc,
Czech Republic
e-mail: pechous@prfnw.upol.cz

generator. Virtual instrument (VI) working as a Mössbauer spectrometer is implemented by LabVIEW 7 Express, and this novel method is based on Waveform Peak Detection (WPkD) function that allows detailed analysis of the acquired signal in energy and time dimensions. This function is a software equivalent of the electronic pulse height analyzer device, so that the optimal treatment of the detector signal from the various detector types is achieved by mathematical processing only.

The purpose of this paper is to report on a new design of Mössbauer spectrometer, where the application of novel programming techniques in Mössbauer spectrometry is used. We experimentally show an increase in the signal/noise ratio by digital signal processing (DSP) before Mössbauer spectrum accumulation. The experimental results show an ability to use various detectors in the transmission and in the backscattering modes, and to tune proper signal acquisition from measuring the pulse height spectra.

## 2 Waveform peak detection and scintillation detectors

The negative pulses from the detector are acquired by a suitable sampling frequency, and locations and magnitudes of their amplitudes can be obtained. In Fig. 1, the signal, acquired from photomultiplier tube with an NaI:Tl scintillation crystal and a $^{57}$Co source, and sampled by 5 MS/s sampling frequency, is depicted. The 8-bit binary representation of the voltage detector signal is used.

The amplitude selection process is controlled by several input parameters of WPkD (see Fig. 2). This function finds the location, magnitude of amplitude and the second derivatives of the peaks in the detector signal. The threshold and width input parameters serve as separation tools of true detector pulses from the noise. The threshold determinates the minimum value of the peak amplitude and the width determinates the minimum peak width according to a number of samples over the threshold.

The amplitude values of the detected peaks, which WPkD generates, are used for the pulse height analysis. The locations output contains positions of valid peaks found in the current block of data. The amplitude values of the detected peaks are used, together with the location values and proper low and high discrimination levels, for an accumulation of Mössbauer spectra. The procedure to set the optimum number of velocity channels was published in [5].

### 2.1 Transmission Mössbauer spectrometry with NaI:Tl and YAP:Ce scintillators

With this Mössbauer spectrometer, a proper use of NaI:Tl and YAP:Ce scintillators, unwinding from different activities of radioactive source, was tested. The negative pulses from these scintillators were acquired by 5 MS/s sampling frequency.

Two Mössbauer spectra of $\alpha$-$^{57}$Fe$_2$O$_3$ (>90% enrichment of $^{57}$Fe) were measured in the same geometry and with the same high voltage on the photomultiplier tube for a time of 1 h. For both detectors, the amplitude analysis was performed for proper setting of the discrimination levels. For NaI:Tl (thickness of 0.15 mm) scintillator, an input range of ±0.25 V with a discrimination window of 33.47 mV was used, and for YAP:Ce (thickness of 0.35 mm) scintillator, an input range of ±0.05 V with an discrimination window of 7.87 mV was used.

If the $^{57}$Co source with an activity, $A$, of 50 mCi has been used, the resonant effect was the same, but the number of counts was twice as high as those observed for YAP:Ce. The same measurements were carried out with a source of 20 mCi activity. In Table 1, the

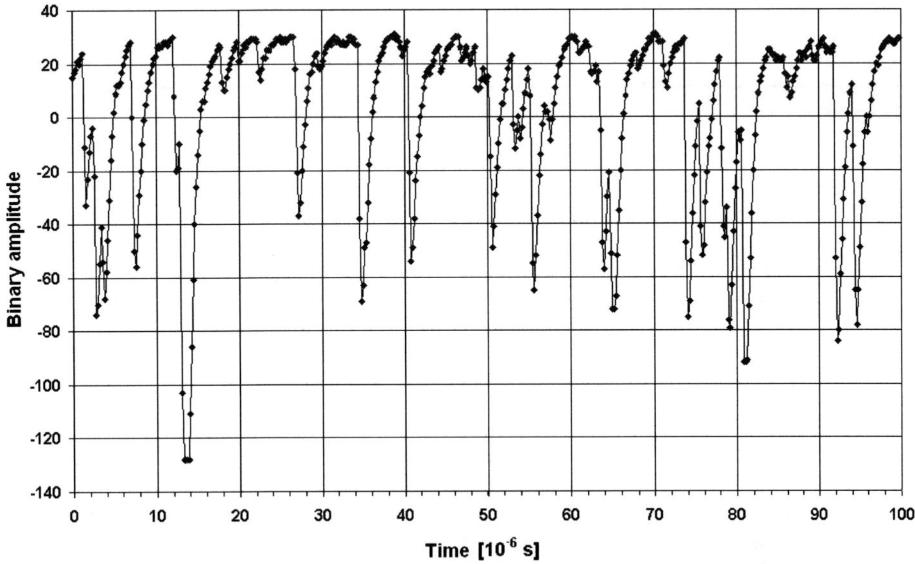

Fig. 1 The NaI:Tl detector signal acquired by 5 MS/s

Fig. 2 Icon of Waveform Peak Detection VI

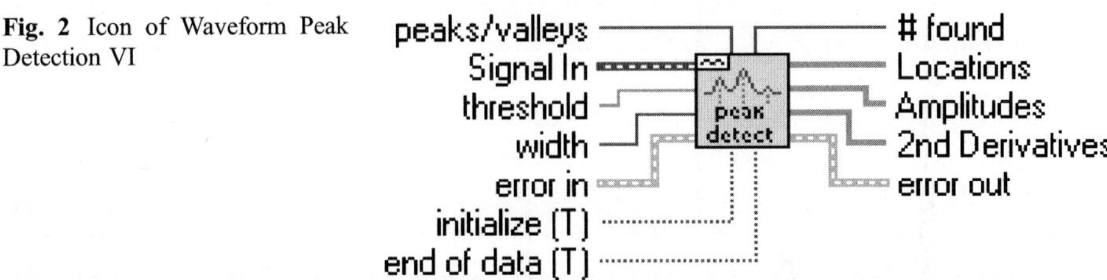


results of these comparing measurements are listed. The productivity, $Q$, was calculated according to the equation in a form of

$$Q = \varepsilon^2 I_{\text{out}}(\infty),$$

where $\varepsilon < 1$ is the resonant effect, and $I_{\text{out}}(\infty)$ is the background intensity in the Mössbauer spectrum [7].

Counts, being twice as high for YAP:Ce scintillator than for NaI:Tl scintillator, confirm that by using high activity sources, NaI:Tl scintillator is overloaded. YAP:Ce is able to registrate a higher activity because of its shorter light scintillation and short-impulse overlapping. From the $Q_{50}/Q_{20}$ ratio, one can see how the productivity of the measurement increases with the high activity source with different scintillators.

2.2 Conversion X-ray Mössbauer spectrometry with NaI:Tl detector

Measuring the conversion X-ray Mössbauer spectrum represented the next motivation for showing the flexibility of the presented system. In this technique, the conversion 6.3 keV X-rays are detected in backscattering geometry. In our configuration, thanks to a high

**Table 1** Results of comparing measurements with NaI:Tl and YAP:Ce scintillators

| Parameter | $A$ [mCi] | NaI:Tl | YAP:Ce |
|---|---|---|---|
| Background | 50 | 143,176 | 284,631 |
|  | 20 | 80,983 | 145,682 |
| Resonant effect [%] | 50 | 26.98 | 27.25 |
|  | 20 | 28.07 | 26.38 |
| Productivity Q | 50 | 10,422 | 21,135 |
|  | 20 | 6,381 | 10,138 |
| Ratio $Q_{50}/Q_{20}$ |  | 1.6 | 2.1 |

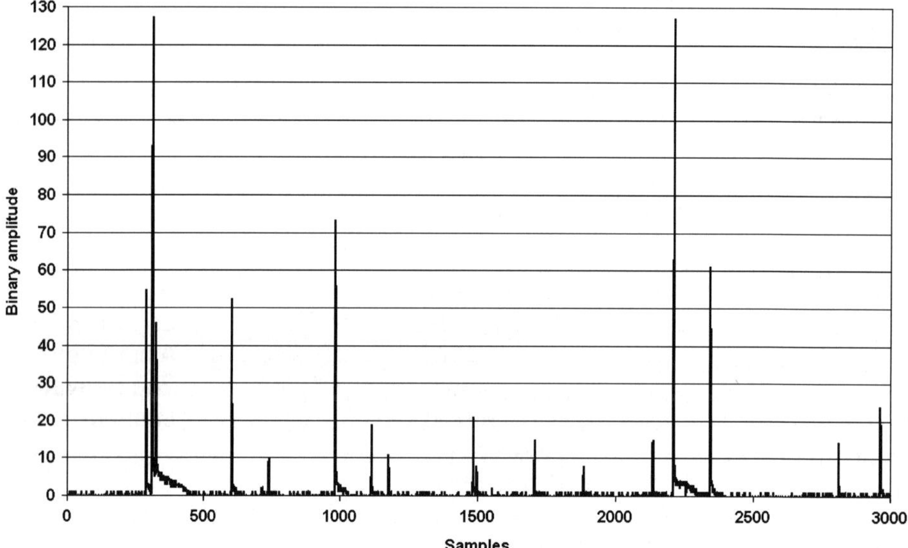

**Fig. 3** Gas-flow proportional counter (CEMS mode) signal

activity source (50 mCi) and thin calibration sample, it was possible to registrate spectrum in the transmission mode. The measurement was performed with NaI:Tl scintillator for 1 h with a resonant effect of 51%.

## 3 WPkD and gas-flow conversion electron detectors

It is also possible to measure conversion electron Mössbauer spectra with presented spectrometer system. In this situation, we used a gas-filled conversion electron counter (90%He + 10%CH$_4$). The pulse-height spectra of the acquired signal from this counter show a typical sharp peak in the number of low energy electrons, but there are counted noise impulses too. The electron energy depends also on the depth from which the electron is emitted.

The typical signal, acquired on the output of this proportional counter, is shown in Fig. 3. This signal was acquired by 1 MS/s sampling frequency, as this value is sufficient

 Springer

for a quality data processing without a loss of any signal information. The acquiring with a higher sampling frequency is possible, but without obtaining any new information on the signal.

In Fig. 3, it is seen that impulses are, in fact, a combination of two components, high-fast (electron) and low-slow (ions) ones.

Furthermore, when applying WPkD in this design, we arrive at the next disadvantage. In Fig. 4a), a detail of the detector signal is shown, and in Fig 4b), amplitudes and locations of founded peaks, declared as a valid by WPkD, are shown. One can see that there are "false" impulses on the bottom of high-energy impulses that can be of higher amplitudes than some impulses for low-energy electrons. It negatively influences the quality of the Mössbauer spectrum when they are counted into the spectrum.

The process of reducing these artificial peaks in the acquired signal is based on the use of the second derivative values (WPkD output) of each founded peak (see the inset of Fig. 4c). In this situation, the magnitude of these "false-peak" second derivatives does not reach a value of $-1$.

The software "noise-reducing" procedure was created as a SubVI (subprogram in main VI), and used before the Mössbauer spectrum accumulation. The effects and results of using this procedure are depicted in Fig. 4d), where only true peaks, belonging to the electron detection are present in the final signal. The $-1$ value as a reducing parameter can be too high sometime and consequently, some true impulses can be deleted.

Various pulse-height spectra of the detector signal, acquired with a different width of WPkD and the second derivative parameters, are shown in Fig. 5. The value of threshold of WPkD was set to 1 in all cases.

In order to choose the best combination of parameters, several Mössbauer spectra, with the same level of background ($I_{out}(\infty) \approx 2.5 \times 10^6$) were measured, see Fig. 6.

The spectra were measured with different width parameter (4, 5, and 6) as an input parameter for WPkD, and different second derivative parameters, used in software procedure before spectra accumulation, in energy window from 5 to 110 channel. Each spectrum was fitted by Recoil 1.0 software to find spectra with the highest statistical quality. The results of the fitting procedures are presented in Table 2, where the values of HWHM (half width at half of maximum) for $w_3$ of the spectral lines 3 and 4 are also listed.

It can be seen that the time of measuring increases with decreasing value of the second derivative. This is in accordance with the shape of the amplitude spectra in Fig. 5 for all widths, where the total number of counts decreases when low-energy impulses are not counted after the reducing process is applied. For reaching the same level of the background (approximately the same statistical quality of the spectrum), we thus need a longer measuring time.

From comparison of $\chi^2$ values ($\chi^2$ represent parameter of correspondence of the fit with Lorentz line shape), we can conclude that with using the second derivative parameter as a noise-reducing tool, $\chi^2$ value decreases with decreasing value of the second derivative.

The Mössbauer spectra were recorded into 750 velocity channels; it means that one channel has the width of about 67 µs. The width of the fast component of the typical impulse in the detector signal is from 20 to 50 µs, and thus, every impulse is correctly recorded into the relevant velocity channel. As the slow component of the impulse can be about 1 ms long and the false impulses generated by WPkD are counted into following channels, their intensities are higher than without this counting, thus increasing their widths (HWHMs). This fact is seen in Table 2, where, based on our noise-reducing process, $w_3$ value decreases, and thus a better resolution of the sub-spectra is possible. However, with

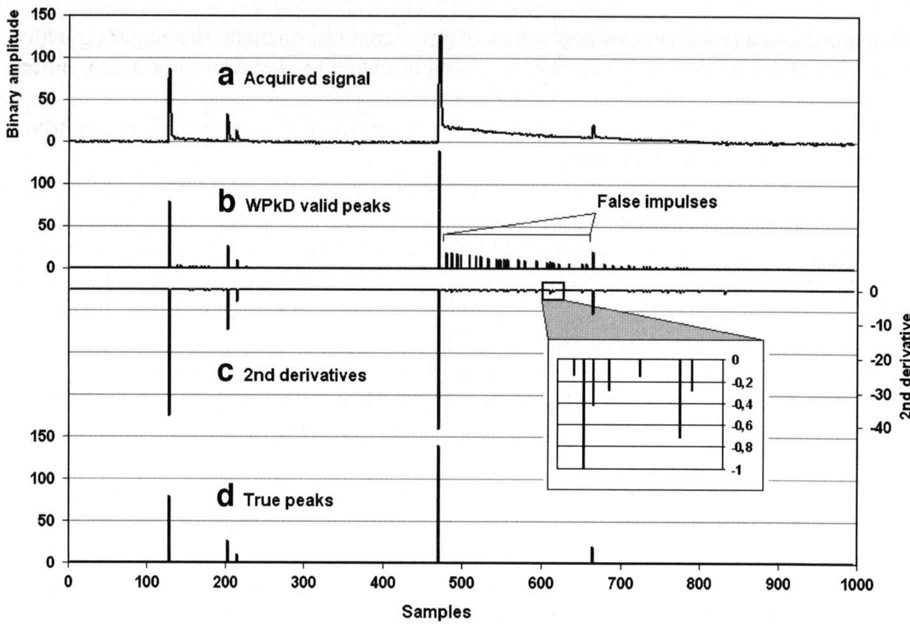



**Fig. 4** Detail of the CEMS signal with overlap impulses

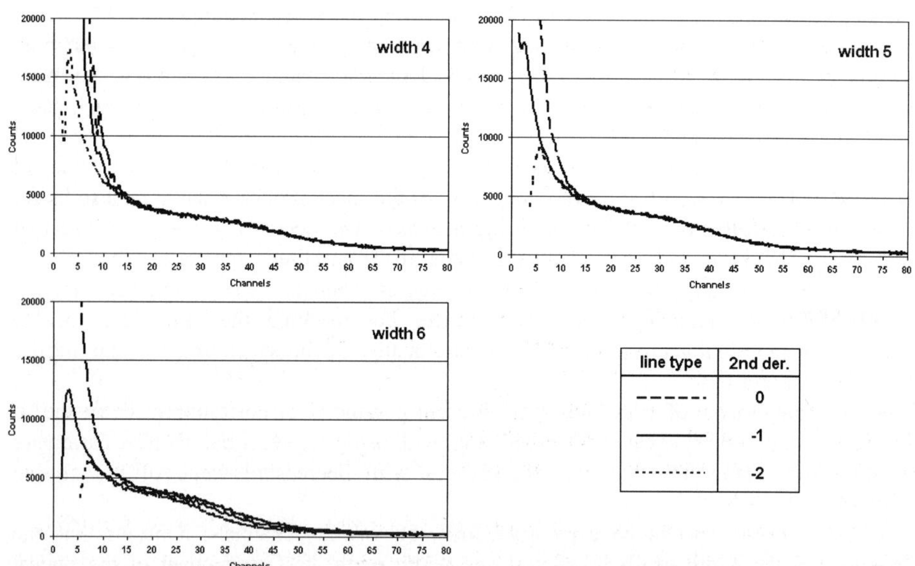



**Fig. 5** Pulse-height spectra for various width and the second derivative parameters

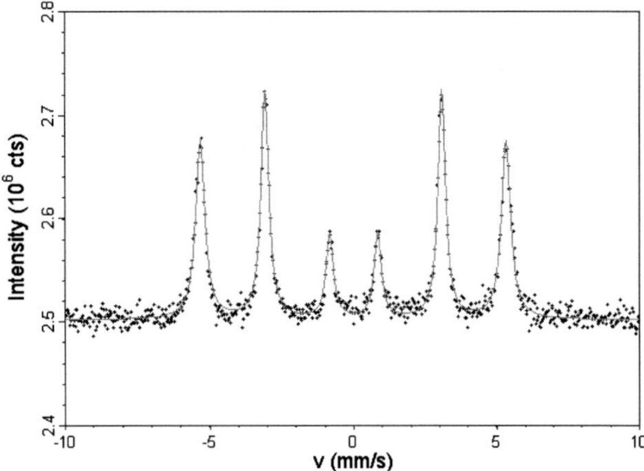

**Fig. 6** Measured Mössbauer spectrum of α-Fe sample

**Table 2** Measurements with various combinations of parameters

| $i$ | Width | Window | 2nd | $\varepsilon$ [%] | $Q$ | Time [h] | $\chi2$ | $w_3$ [mm/s] |
|---|---|---|---|---|---|---|---|---|
| 1 | 4 | 5–110 | 0 | 8.4 | 17,693 | 34 | 59.1 | 0.21 |
| 2 | 4 | 5–110 | −1 | 7.1 | 12,626 | 44 | 25.7 | 0.15 |
| 3 | 4 | 5–110 | −2 | 7.3 | 13,283 | 51 | 19.1 | 0.16 |
| 4 | 5 | 5–110 | 0 | 7.6 | 14,451 | 43 | 27.9 | 0.16 |
| 5 | 5 | 5–110 | −1 | 7.6 | 14,632 | 44 | 19.2 | 0.14 |
| 6 | 5 | 5–110 | −2 | 7.3 | 13,271 | 65 | 20.1 | 0.15 |
| 7 | 6 | 5–110 | 0 | 7.7 | 14,920 | 46 | 24.7 | 0.16 |
| 8 | 6 | 5–110 | −1 | 7.7 | 14,857 | 50 | 18.6 | 0.13 |
| 9 | 6 | 5–110 | −2 | 7.8 | 15,236 | 54 | 19.5 | 0.15 |

too high value of the reducing parameter (−2 in this case), $\chi^2$ values and widths $w_3$ increases and useful signal is deleted.

## 4 Conclusion

The novel design, significantly improving the efficiency of the Mössbauer measurements and the precision of their results, was constructed. Virtual Mössbauer spectrometer, based on LabVIEW graphical programming environment and Waveform peak detection function, is new tool in Mössbauer spectroscopy and offers various combinations of detectors and measurement geometries.

The use of this novel embodiment of Mössbauer spectrometer, based on WPkD, offers a further combination of input parameters for a better signal acquisition, actually depending on the situation. One can, for instance, use a higher sampling frequency for faster detectors, which involve number of samples in one impulse in acquired signal, or change the width and threshold as a WPkD input parameters for moving the level for noise reduction in the signal. It is possible to registrate more quality spectra with a higher sampling frequency. However, one has to bear in mind that this results in too much data for DSP analyzing and therefore, a higher computation time.

 Springer

Disadvantage of this solution lies on the fact that the performance of such a spectrometer is given by the performance of the computer used (PC or PXI). Detecting the gamma-ray with this system, there are no evident troubles, however, when used for detecting the conversion electrons, there are naturally false impulses generated and then other DSP routine is necessary to be developed.

**Acknowledgements** Financial support from The Ministry of Education, Youth and Sports under project of research center No 1M6198959201 is gratefully acknowledged.

# References


1. Mashlan, M., Jancik, D., Zak, D., Dufka, F., Snasel, V., Kholmetskii, A.L.: The Mössbauer spectrometer as a virtual instrument. In: Miglierini, M., Petridis, D. (eds.) Mössbauer spectroscopy in materials science, pp. 399–406. Kluwer, Boston, MA (1999)
2. Morales, A.L., Zuluaga, J., Cely, A., Tobon, J.: Autonomous system design for Mössbauer spectra acquisition. Hyperfine Interact. **134**, 167–170 (2001)
3. Velasques, A.A., Trujillo, J.M., Morales, A.L., Tobon, J.E., Reyes, L., Gancedo, J.R.: Design and construction of an autonomous control system for Mössbauer Spectrometry. Hyperfine Interact. **161**, 139–145 (2005)
4. *LabVIEW 7 Express, User manual*. National Instruments. National Instruments Corp. (2003)
5. Pechousek, J., Mashlan, M.: Mössbauer spectrometer in the PXI/CompactPCI modular system. Czech. J. Phys. **55**, (7), 853–863 (2005)
6. *The Measurement and Automation Catalog 2006*. National Instruments, National Instruments Corp. (2005)
7. Kholmetskii, A.L., Mashlan, M., Nomura, K., Misevich, O.V., Lopatik, A.R.: Fast detectors for Mössbauer spectroscopy. Czech. J. Phys. **52**, (7), 763–771 (2001)

Hyperfine Interact (2007) 175:9–14
DOI 10.1007/s10751-008-9581-4

# Room temperature $^{57}$Fe Mössbauer spectroscopy of ordinary chondrites from the Atacama Desert (Chile): constraining the weathering processes on desert meteorites

M. Valenzuela · Y. Abdu · R. B. Scorzelli · M. Duttine ·
D. Morata · P. Munayco

Published online: 24 April 2008
© Springer Science + Business Media B.V. 2008

**Abstract** We report the results of a study on the weathering products of 21 meteorites found in the Atacama Desert (Chile) using room temperature $^{57}$Fe Mössbauer spectroscopy (MS). The meteorites are weathered ordinary chondrites (OCs) with unknown terrestrial ages and include the three chemical groups (H, L, and LL). We obtained the percentage of all the Fe-bearing phases for the primary minerals: olivine, pyroxene, troilite and Fe–Ni metal, and for the ferric alteration products (composed of the paramagnetic $Fe^{3+}$ component and the magnetically ordered $Fe^{3+}$ components) which gives the percentage of oxidation of the samples. From the Mössbauer absorption areas of these oxides, the terrestrial oxidation of the Atacama OC was found in the range from ~5% to ~60%. The amount of silicates as well as the opaques decreases at a constant rate with increasing oxidation level.

**Keywords** Weathered ordinary chondrites · Iron oxi-hydroxides · Hot desert environments ·
Atacama Desert

## 1 Introduction

Ordinary chondrites (OCs), one of the most abundant type of known meteorites (~80%), are excellent standard geological samples to track the effects of terrestrial weathering because: (1) their initial composition before weathering is very well known from the analysis of modern falls [1–4], (2) all the iron in an equilibrated OCs is considered to be present as $Fe^{2+}$ or $Fe^0$ (in silicates and sulphides in the first case, and Fe–Ni metal in the second), thus any

M. Valenzuela (✉) · D. Morata
Departamento de Geología, Universidad de Chile, Plaza Ercilla #803, Santiago, Chile
e-mail: edvalenz@cec.uchile.cl

Y. Abdu · R. B. Scorzelli (✉) · M. Duttine · P. Munayco
Centro Brasileiro de Pesquisas Físicas (CBPF/MCT), Xavier Sigaud 150, Río de Janeiro, Brazil
e-mail: scorza@cbpf.br

$Fe^{3+}$ present in an ordinary chondrite find (as opposed to an observed fall) may be interpreted as the product of terrestrial alteration that can be quantified by $^{57}Fe$ Mössbauer spectroscopy, (3) their terrestrial ages (i.e., the time they have spent on Earth since falling) can be constrained through $^{14}C$ [5], $^{81}Kr$, $^{36}Cl$ [6] or $^{26}Al$ analyses to provide a chronology of events, and (4) because these meteorites fall more or less uniformly over the Earth's surface, they provide a standard sample to study weathering processes in very different environments.

The aim of this study is the identification and quantification of the weathering products of 21 weathered OCs found in two main areas of the Atacama Desert, northern Chile, one of the oldest and driest deserts in the world [7, 8], in order to understand the weathering processes operating in the Atacama Desert (AD), and to constrain the effect of the terrestrial environment on the primary signatures of this extraterrestrial material.

This study is being complemented with other techniques such as optical and scanning electron microscopy (SEM), X ray diffraction (XRD), physical and magnetic properties characterization, $^{14}C$ dating, and bulk and mineral grain chemistry of major and trace elements.

## 2 Background

Almost 90% of the normative mineralogy of ordinary chondrites is made up of olivine, pyroxene, troilite and Fe–Ni alloys, all of which contain Fe. The alteration products that have been recognized in desert OCs are mainly iron oxi/hydroxides (e.g., akaganéite, goethite, maghemite and magnetite) formed by the oxidation of the primary phases. As Mössbauer spectroscopy is extremely sensitive to changes in the Fe valence state, it can provide an overview of the effect of weathering on the whole sample and quantify the relative proportion of ferrous, ferric and metallic components, allowing the quantification of the oxidation of the samples.

## 3 Experimental

About 1–2 g of sample taken from the outer part of the meteorites (but excluding any surviving fusion crust) were crushed to obtain powders for $^{57}Fe$ Mössbauer spectroscopy and XRD.

Mössbauer spectra were recorded at room temperature (RT), in high velocity, in transmission geometry using a $^{57}Co(Rh)$ source, in a Halder spectrometer with 512 channels. The velocity of the drive shaft tracks the input waveform to the drive control unit. With a sinusoidal wave input, the shaft velocity changes uniformly in time, that is, the acceleration is constant and a velocity reference signal positive. The drive velocity was calibrated with the same source and a metallic iron foil at RT. Average recording time was 48 h per sample. Mössbauer absorbers containing ~250 mg of the bulk meteorite powdered sample were used. Isomer shifts were measured relative to α-Fe. NORMOS code [9] was used for the spectrum analysis.

The phase quantification was done using the relative areas taken from fitted Mössbauer spectrum, since they are proportional to the number of resonant nucleus that represents the Mössbauer effect occurrence probability. It should be mentioned that, for a given temperature, this probability is not equal for all the phases. However, in this paper we

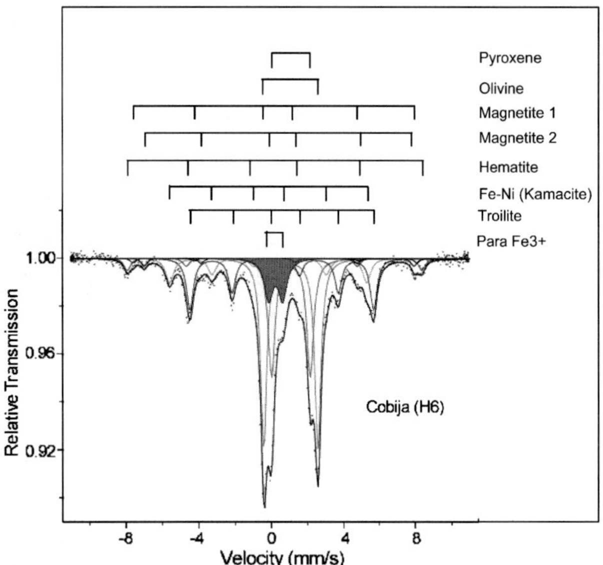



**Fig. 1** RT Mössbauer spectrum of the Cobija (*H6*) meteorite

are considering as a good approximation, that all the phases have the same probability according to reported low temperature results in similar samples.

## 4 Results and discussion

The first petrographic studies with traditional microscopy and SEM showed a wide range of weathering grades for the set of samples given by different patterns of alteration, mainly of the opaque phases, and also in shock veins [10]. From XRD, the alteration products detected were akaganéite, goethite, maghemite, hematite and magnetite [10, 11].

The relative proportions of the major Fe-bearing phases of the samples (as pyroxene, olivine, Fe–Ni and troilite), as well as the alteration products, were recognized and quantified by Mössbauer spectroscopy. As can be seen in Fig. 1, the Mössbauer spectrum shows an overlapping of paramagnetic and magnetic phases. The spectra were fitted with three quadrupole doublets, two of them attributed to $Fe^{2+}$, one associated to olivine (IS= 1.19 mm/s; QS=3.0 mm/s) and the other to pyroxene (IS=1.21 mm/s; QS=2.14 mm/s). The third is due to $Fe^{3+}$ (IS=0.36 mm/s; QS=0.77 mm/s) which can be associated to superparamagnetic oxides and/or iron hydroxides (small particles of goethite, akaganéite, lepidocrocite), that will be identified by low temperature measurements, under way. Finally, the magnetic components have been associated to large-particle goethite ($B_{hf}$=38.2 T), Fe–Ni ($B_{hf}$=33.5 T), troilite ($B_{hf}$=32.5 T), hematite ($B_{hf}$=51.8 T) and magnetite (two sextets $B_{hf}$=49.1 T and $B_{hf}$=46.2 T).

The samples show a wide range in total oxidation (between 4.9% and 62.8%), given by the presence of Fe-oxides appearing as magnetically ordered $Fe^{3+}$ or as paramagnetic $Fe^{3+}$.

To study the weathering effect on the primary mineralogy of the samples we compare the spectral areas of ferromagnesian silicates (olivine and pyroxenes) and opaque (Fe–Ni and

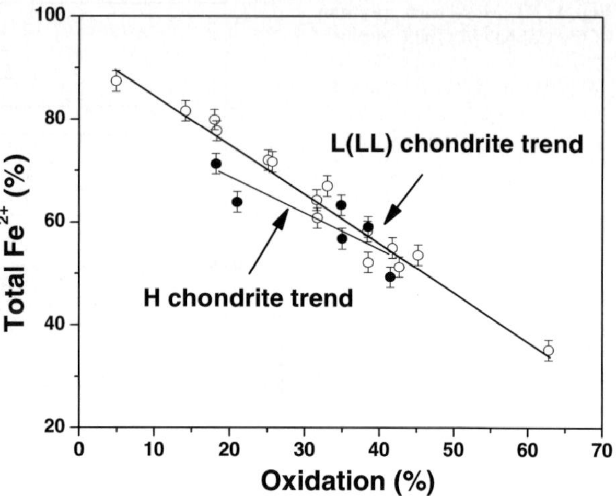

**Fig. 2** Spectral area of total $Fe^{2+}$ (%) against total ferric oxidation derived from Mössbauer spectra recorded from weathered Atacama OCs, showing that ferromagnesian silicates are weathered almost at a constant rate



troilite) with the total amount of oxidation to determine which phases are most susceptible to weathering (Figs. 2 and 3). We can observe a decrease in primary phases with increasing oxidation, almost at a constant rate for the H and L–LL groups, suggesting that all iron-containing minerals within the meteorite are affected by weathering to some degree.

From the Mössbauer data we plot the percentage oxidation of the samples as a frequency histogram. Comparing this oxidation frequency distribution of the Atacama Desert (AD) OCs (Fig. 4c) with the oxidation frequency distribution of other hot desert OCs [12], from the Nullarbor Region (NR) and Roosevelt County (RC; Fig. 4a) and from the Sahara Desert (Fig. 4b) we can observe that AD OCs show a peak around 35%, similar to Sahara Desert samples, in contrast with the peak value between 40–45% of NR and RC samples.

The authors of Ref. [13] related this peak to the level of weathering required before meteorites begin to be eroded and lost from a population. As this property is related to the physical weathering of OCs, in principle it should be the same for all collection areas. Therefore, the difference we observe can be related to the stability and age of the accumulation surfaces. AD OCs display a distribution pattern similar to SD OCs that corresponds to a younger accumulation area compared with NR and RC [12].

## 5 Conclusions

The decrease in primary phases with increasing oxidation suggests that all iron-containing minerals within the meteorite are affected by weathering to some degree, as previously reported for OCs from other hot deserts [12].

Although we have not yet obtained the terrestrial ages of these meteorites we can suppose from the oxidation frequency distribution pattern that the main accumulation area, in which most of the samples have been found, can not be as old as the time required to reach the peak value of about 40–45%, found for other hot desert meteorites [12, 13].

 Springer

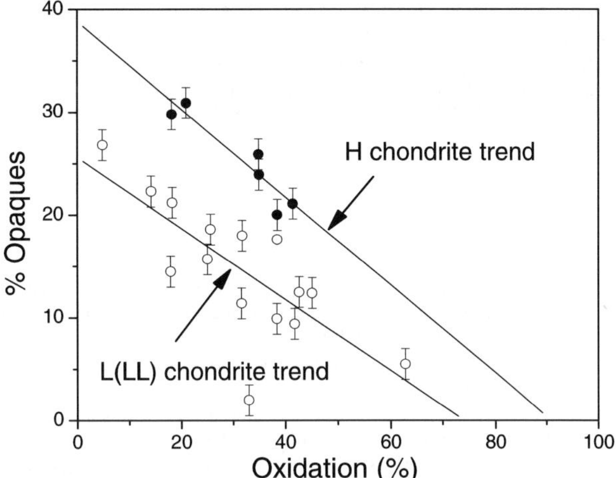

**Fig. 3** Spectral area of opaques (Fe–Ni and troilite) against total ferric oxidation, showing that both groups of Atacama OCs are affected at almost constant rate by weathering



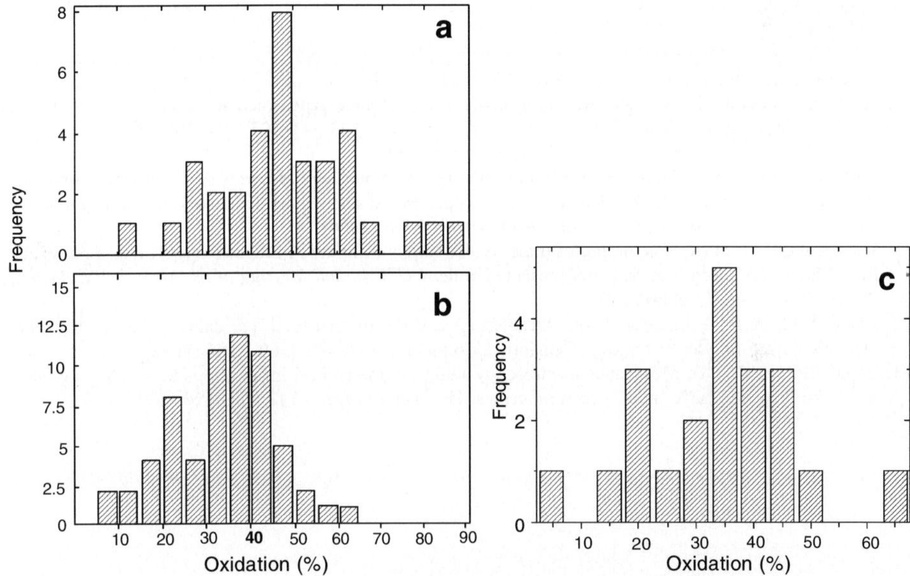



**Fig. 4** Percentage oxidation (in 5% bins) against frequency for all the hot desert OCs studied by [12], split between (a) non-Sahara Desert (i.e., Roosevelt County and Nullarbor Region) samples and (b) Sahara Desert samples. **c** Shows the oxidation-frequency distribution for Atacama Desert (AD) samples obtained in the present study

As pointed out by [12], at least two factors are significant in determining the level of oxidation: terrestrial age and initial chemistry. H OCs should show, for a given terrestrial age, higher oxidation levels than L(LL) OCs, because of their high Fe–Ni content, so that the next step of terrestrial age dating will be crucial in understanding the weathering processes in the Atacama Desert.

**Acknowledgments** Thanks to Rodrigo Martínez, Edmundo Martínez and Enrique Stucken for the meteorite sample donations and to the Natural History Museum of London and Museo Nacional de Historia Natural de Chile for other sample loans. We are also grateful to the Servicio Nacional de Geología y Minería de Chile and Edivaldo S. Filho from CBPF. Special thanks to CONICYT for a research stage grant at the CBPF, as well as CONICYT-BIRF for a grant supporting the doctoral studies of Millarca Valenzuela (2004–2007).

# References


1. Jarosewich, E.: Chemical analyses of meteorites: a compilation of stony and iron meteorite analyses. Meteoritics **25**, 323–337 (1990)
2. Clayton, R.N., et al.: Oxygen isotope studies of ordinary chondrites. Geochim. et Cosmochim. Acta **55**, 2317–2337 (1991)
3. McSween, H.Y., et al.: The mineralogy of ordinary chondrites and implications for asteroid spectrophotometry. Icarus **90**, 107–116 (1991)
4. Kallemeyn, G.W., et al.: Ordinary chondrites: bulk compositions, classifications, lithophile-element fractionations, and composition-petrographic type relationships. Geochim. et Cosmochim. Acta **53**, 2747–2767 (1989)
5. Jull, A.J.T., et al.: Carbon-14 activities in recently fallen meteorites and Antarctic meteorites. Geochim. et Cosmochim. Acta **53**, 2095–2100 (1989)
6. Nishiizumi, K., et al.: Update on terrestrial ages of Antarctic meteorites. Earth Planet. Sci. Lett. **93**, 299–313 (1989)
7. Dunai, T.J., et al.: Oligocene–Miocene age of aridity in the Atacama Desert revealed by exposure dating of erosion-sensitive landforms. Geology **33**, 321–324 (2005)
8. Arancibia, G., et al.: K–Ar and $^{40}$Ar/$^{39}$Ar geochronology of supergene processes in the Atacama Desert, Northern Chile: tectonic and climatic relations. J. Geol. Soc. **163**, (1), 107–118 (2006)
9. Brand, R. A.: NORMOS, Mössbauer fitting program (1992)
10. Valenzuela, M. et al.: Estudio de la meteorización de tres condritos ordinarios del Desierto de Atacama (N. Chile) a través de MEB, DRX y espectroscopía Mössbauer. Paper presented at the *XI Congreso Geológico Chileno*, Antofagasta, Chile, 7–11 August 2006 (2006)
11. Valenzuela, E.M. et al.: Weathering of ordinary chondrites from the Atacama Desert, Chile: first results from Mössbauer spectroscopy. Paper presented at the *69TH Annual Meeting of the Meteoritical Society*, Zürich, 6–11 August 2006 (2006)
12. Bland, P.A., et al.: Climate and rock weathering: a study of terrestrial age dated ordinary chondritic meteorites from hot desert regions. Geochim. Cosmochim. Acta **62**, 3169–3184 (1998)
13. Bland, P.A., et al.: $^{57}$Fe Mössbauer spectroscopy studies of meteorites: implications for weathering rates, meteorite flux, and early solar system processes. Hyperfine interact. **142**, 481–494 (2003)

Hyperfine Interact (2007) 175:15–21
DOI 10.1007/s10751-008-9582-3

# The metallurgic furnaces at the Curamba Inca site (Peru): a study by Mössbauer spectroscopy and X-ray diffractometry

Yezeña Huaypar · Luisa Vetter · Jorge Bravo

Published online: 20 March 2008
© Springer Science + Business Media B.V. 2008

**Abstract** The Inca site at Curamba is located in the Province of Apurimac in the southern highlands of Peru where, according to some historians, several thousand furnaces used for ore smelting were found. For this work, four samples of burned soil were gathered from these furnaces and classified as Curamba1, Curamba2, Curamba3, and Curamba4, and studied using transmission Mössbauer spectroscopy (TMS) and X-ray diffractometry (XRD). The mineralogical composition of the samples was determined by XRD and the structural sites in the minerals occupied by iron cations were characterized by TMS. Moreover, an attempt was made to determine the maximum temperature reached in these furnaces using the refiring technique of the samples in an oxidizing environment and monitoring the structural modifications at the iron sites by changes in the Mössbauer hyperfine parameters. The TMS results of Curamba2 show that the maximum temperature reached in this furnace was about 900°C, in agreement with the mineralogical composition found by XRD. In the case of Curamba1 and Curamba4 the maximum temperature estimated was about 400°C.

**Keywords** Furnaces · Smelting · Refiring · Oxidizing environment ·
Mössbauer spectroscopy · X-ray diffractometry

## 1 Introduction

This work is intended to test the usefulness of $^{57}$Fe Mössbauer spectroscopy (MS) and X-ray diffractometry (XRD) to obtain information about metallurgic furnaces and the maximum

Y. Huaypar (✉) · J. Bravo
Facultad de Ciencias Físicas, Universidad Nacional Mayor de San Marcos, Av. Venezuela Cdra. 34 s/n,
Lima 01, Perú
e-mail: yhuaypar@yahoo.es

J. Bravo
e-mail: jbravoc@unmsm.edu.pe

L. Vetter
Pontificia Universidad Católica del Perú, Av. Universitaria 1801, Lima 32, Perú
e-mail: luchivetter@hotmail.com

temperature reached with them by comparing the results of scientific investigation with the available historical information.

As a technique to study archaeological clay-based materials, MS is unique since it provides information about the clay phases that contain iron and their transformations during thermal treatment. Several efforts have shown that the changes observed in the Mössbauer spectra, of clay-based materials, caused by thermal treatment at temperatures in the range of 400°C to 600°C is due to the loss of hydroxyl groups in the clay minerals with the consequent distortion of the lattice and increase in the electric quadrupole splitting (QS) at the structural $Fe^{3+}$ sites in the clay minerals. At higher temperatures, near 800°C, the laminar structure eventually collapses into an amorphous phase and the QS starts to decrease. Finally new crystalline phases are formed. This behaviour allows transmission Mössbauer spectroscopy (TMS) to be used as a thermometer to determine the firing temperature of ancient materials [1].

On the other hand, XRD is an important technique in mineralogy for the identification, quantification, and characterization of minerals in complex mineral mixtures, which complement the information obtained by TMS.

The refiring technique assumes that the mineral composition of the as found clay-based material has remained unaltered since its original firing; this mineral composition is related to a set of characteristic hyperfine parameters. On refiring, these hyperfine parameters are not supposed to change until the original firing temperature is surpassed, at which point the mineral composition of the material resumes its evolution under the thermal treatment [2–4].

Curamba is an archaeological site with Inca occupation, located at 3,646 m above sea level in the Department of Apurimac and 28 km from the City of Abancay. It is on a plateau and exposed to strong winds. The furnaces are distributed over terraces. The remains of Curamba are one of the outstanding testimonies of religious architecture represented by the "Ushnu" or pyramidal structures and of its strategic defensive placement, in addition to the Inca metallurgical technology represented by the furnaces or "huayras", which are the subject of this study [5].

According to historical documents recompiled by Olaechea [6], there were more than 40 chambers and groups of furnaces distributed in groups of three. The furnace in the middle was of rectangular shape, long and narrow, measuring about $3 \times 0,7 \times 0,7$ m; the other two were of similar size but ended in an oval space, of dome shape, open at the top. One end of this structure coincided with the vertical face of the terrace so that air may enter the structure (Fig. 1). According to in situ observations, these furnaces were constructed from polished limestone and mud as mortar.


## 2 Experimental procedures

The samples were collected from four furnaces. Each of these samples was in the form of lumps of about 50 g that were picked up from the interior of the furnaces and classified as Curamba1, Curamba2, Curamba3, and Curamba4. The fired lumps constituted hard aggregates, resistant to pressure with the fingers and reddish coloured. (Fig. 2). The refiring of the samples was done in the temperature range from 400°C to 1,000°C in steps of 100°C, for 18 h per step, in an oxidizing atmosphere due to the circulation of air [5]. Sample aliquots of about 1 g were placed on briquettes resistant to high temperatures and placed inside the furnace quartz tube with the furnace at room temperature (RT) and then heated to the required temperature in 2 h. At the end of the refiring period the furnace was turned off

**Fig. 1** Furnace distribution
according to Olaechea

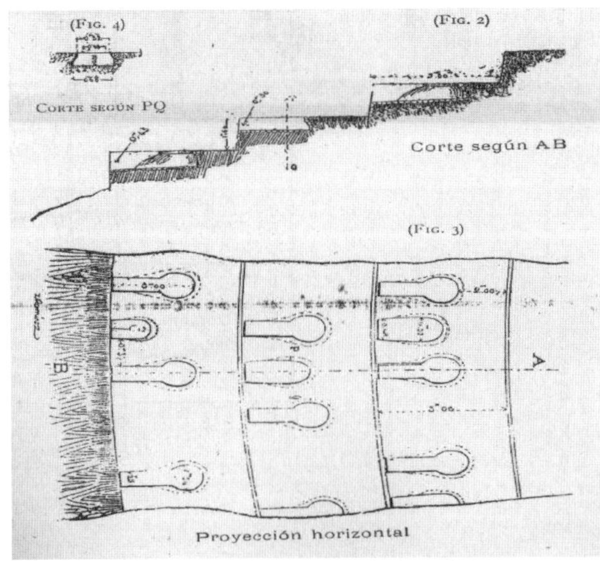


**Fig. 2** Lump from the furnace

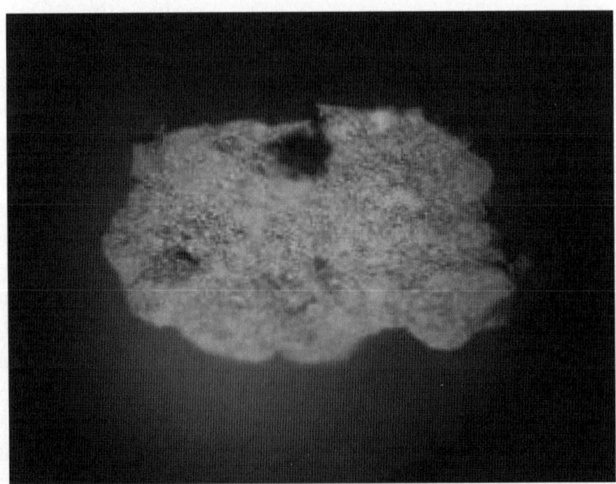

and the samples were allowed to cool down with the furnace, within about 1 h. For each refiring temperature a different sample aliquot was used.

## 2.1 Experimental conditions for MS and XRD measurements

The Mössbauer spectra were obtained with aliquots of 250 mg, for both the as found and refired aliquots. A conventional transmission spectrometer was used, with a $^{57}$Co radioactive source in a Rh matrix. All the spectra were taken at RT.

For the XRD analysis a RIGAKU, model Miniflex, diffractometer was used, with Ni filtered Cu-K$\alpha$ radiation. The experimental setup used a 2$\theta$ interval from 5 to 70°, with a step of 0.02°, and a counting interval of 2 s/step. A powder mount was used, requiring about 500 mg of material [7]. Only the as found samples were analyzed by XRD.

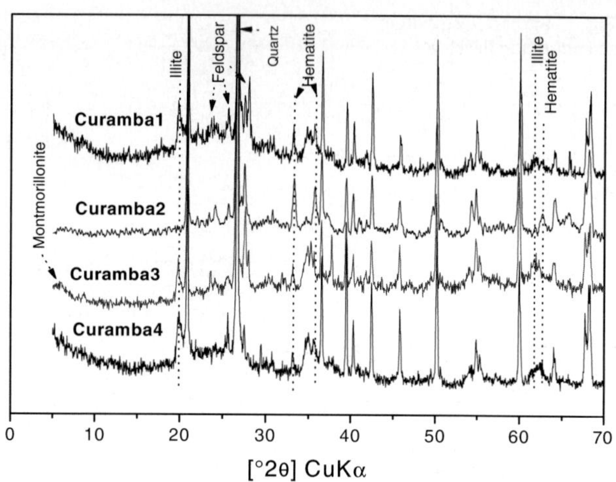



**Fig. 3** X-ray diffraction patterns of Curamba samples

## 3 Results and discussion

The results of the mineralogical analyses of the as found samples by XRD show that quartz, albite, and orthoclase are the major mineral components, and that hematite, illite, and montmorillonite are minor components (Fig. 3). The Curamba3 sample is the one that presents the highest content of clay minerals; from this fact it is deduced that this sample was the least affected by the original firing. For the Curamba2 sample the opposite is observed, in the sense that not a single clay mineral phase is observed in its composition, it is the sample with the highest content of hematite; this fact suggests that its original firing temperature was higher than 850°C which is the decomposition temperature for illite. In the case of samples Curamba1 and Curamba4, the presence of illite in their composition suggests that their original firing temperature was not higher than 850°C, since above this temperature illite decomposes or collapses structurally.

The results of the TMS analyses of the as found Curamba1 and Curamba4 samples (Fig. 4) show two magnetic Fe sites (45%), a paramagnetic $Fe^{3+}$ site (48%) and an undefined Fe site (7%) giving rise to a very broad unresolved singlet. The magnetic site with the higher hyperfine magnetic field of about 51 T is attributable to hematite and the second magnetic site may be associated with a poorly crystallized hematite with cationic substitution. For the as found Curamba2, the spectrum is simpler since only two sites are required, a magnetic (68%) one attributable to hematite and a paramagnetic one (32%). Table 1 lists the more relevant hyperfine parameters for the as found and refired samples measured by TMS.

Figure 5 shows the curves that represent the variation of the quadrupole splitting of paramagnetic $Fe^{3+}$ with the refiring temperature for samples Curamba1, Curamba2 and Curamba4.

## 4 Conclusions

Considering the results obtained, it can be said with certainty that the furnaces did produce high temperatures and that they had actually been in use.

 Springer

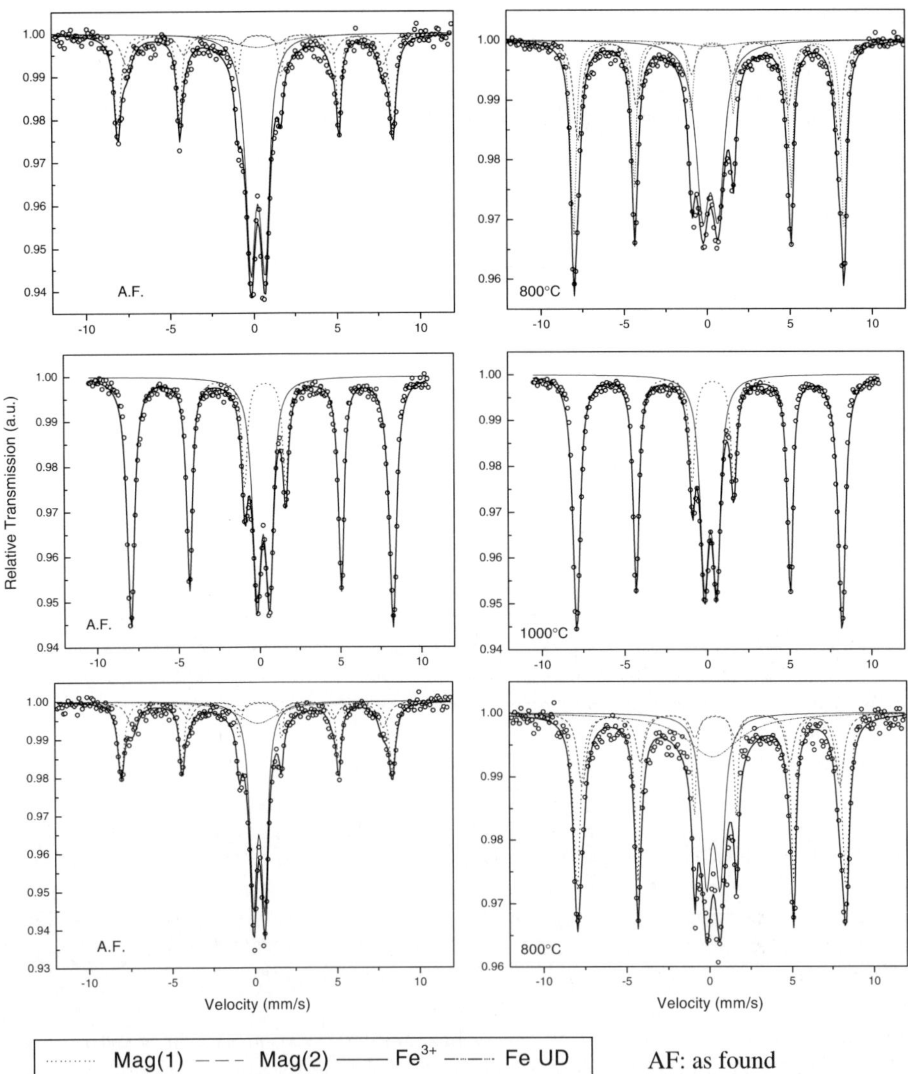



**Fig. 4** Mössbauer spectra taken at RT of as found (*left*) and refired (*right*) Curamba1 (*top*), Curamba2 (*middle*) and Curamba4 (*bottom*) samples

**Table 1** Characteristic Mössbauer parameters for as found Curamba1, Curamba2 and Curamba4 samples measured at RT without refiring

| Sample | H(1)-Mag | IS(1)-Mag | QS(1)-Mag | A(1)-Mag | H(2)-Mag | IS(2)-Mag | QS(2)-Mag | A(2)-Mag | Q-Fe$^{3+}$ | IS-Fe$^{3+}$ | A-Fe$^{3+}$ |
|---|---|---|---|---|---|---|---|---|---|---|---|
| Curamba1 | 51.1 | 0.258 | −0.214 | 0.22 | 48.16 | 0.224 | −0.213 | 0.23 | 0.853 | 0.242 | 0.48 |
| Curamba2 | 50.3 | 0.259 | −0.218 | 0.68 | | | | | 0.753 | 0.206 | 0.32 |
| Curamba4 | 50.8 | 0.262 | −0.209 | 0.24 | 47.11 | 0.31 | −0.16 | 0.23 | 0.687 | 0.248 | 0.42 |

The hyperfine fields are given in Tesla, the IS (relative to $^{57}$Co:Rh) and the QS in millimeter per second

**Fig. 5** Dependence of the QS, isomer shift (*IS*) and relative area of the paramagnetic $Fe^{3+}$, and of the relative areas of the magnetic and undefined iron site of samples Curamba1 (*top*), Curamba2 (*middle*) and Curamba4 (*bottom*)

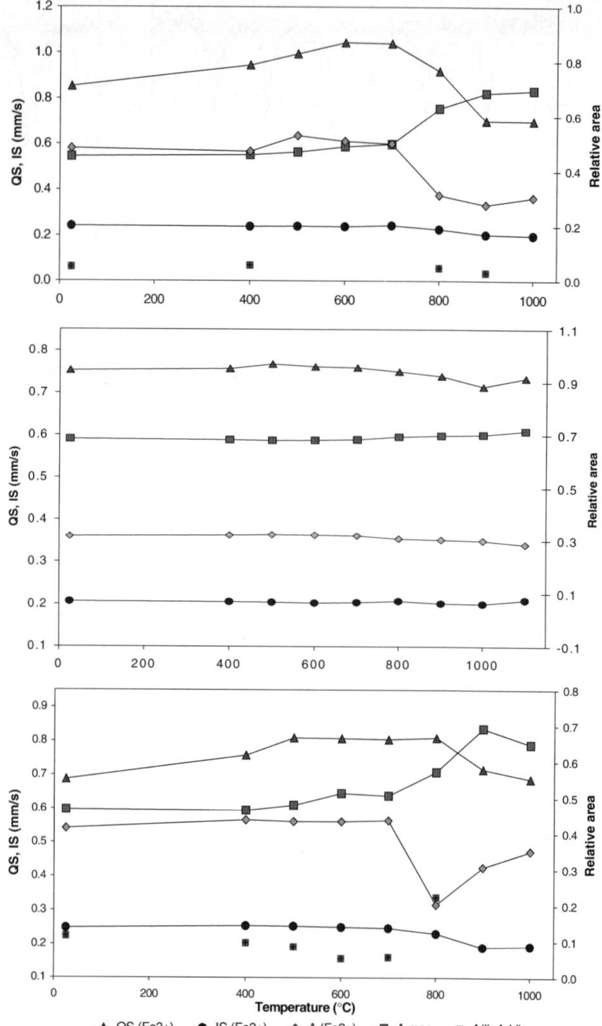


The observed variation of the quadrupole splitting of paramagnetic $Fe^{3+}$ with the refiring temperature implies that the firing temperature in antiquity did not surpass 400°C in samples Curamba1 and Curamba4, since it is unlikely that rehydroxylation may have taken place after firing at a temperature higher than 400°C.

In the case of the sample Curamba2 one can infer that the estimated firing temperature in antiquity was about 900°C. This result suggests that the furnaces may have been used for the preparation of silver–copper alloys, with fusion temperatures near 780°C, which agrees with the documentation recompiled by the archaeologists about the use of these furnaces.

**Acknowledgements** We thank Drs. Fritz and Ursel Wagner, Technische Universität München, for valuable discussions, and the XRD and MS Laboratories at San Marcos University for their services.

Springer

# References


1. Wagner, F.E., Kyek, A.: Mössbauer spectroscopy in archaeology: introduction and experimental considerations. Hyperfine Interact. **154**, 5–33 (2004)
2. Wagner, F.E., Wagner, U.: Mössbauer spectra of clays and ceramics. Hyperfine Interact. **154**, 35–82 (2004)
3. Shimada, I., Häusler, W., Hutzelmann, T., Wagner, U.: Early pottery making in coastal Peru. Part I: Mössbauer study of clays. Hyperfine Interact. **150**, 73–89 (2003)
4. Shimada, I., Merkel, J.: Copper-alloy metallurgy in ancient Perú. Scientific American July, 80–86 (1991)
5. Amorín, J., Alarcón, E.: Prospección arqueológica en el sitio Inca de Curamba y su relación con Sondor. In: XII Congreso Peruano del hombre y la Cultura Andina, pp. 287–294. Universidad Nacional de San Cristóbal de Huamanga, Ayacucho (1999)
6. Olaechea, T.: Apuntes sobre el Castillo y fundición de Curamba. In: Anales de Construcción Civiles, Minas e Industrias del Perú, vol 1, pp. 1–21. Escuela de Ingenieros de Lima, Lima (1901)
7. Moore, D.M., Reynolds Jr., R.C.: X-Ray diffraction and the identification and analysis of clay materials, 2nd edn. Oxford University Press, Oxford (1997)

Hyperfine Interact (2007) 175:23–28
DOI 10.1007/s10751-008-9583-2

# Study of the structural modifications in activated clays by Mössbauer spectroscopy and X-ray diffractometry

Yezeña Huaypar · Jorge Bravo · Abel Gutarra ·
Erika Gabriel

Published online: 20 March 2008
© Springer Science + Business Media B.V. 2008

**Abstract** In this work we study the changes induced on the structure of a smectite clay by chemical acid activation with HCl using X-ray diffractometry (XRD) and transmission Mössbauer spectroscopy (TMS) techniques. By XRD we were able to determine the mineralogical composition of the clay samples and measure the changes in the interplanar distance associated to the structural modifications in the clays. We measured a reduction in the interplanar distance and reflection intensity as the acid concentration in the activation process increased. TMS allowed us identify and characterize the structural sites occupied by ferric and ferrous iron cations. In addition, we were able to monitor the effects caused by the chemical acid activation on the valence state of the iron cations that occupy these structural sites in the clay. For the treatment at low acid concentration, keeping time and temperature of activation constant, our results showed a strong effect on the ferrous and ferric iron sites, reducing and increasing their adsorption relative areas respectively.

**Keywords** Smectite · Chemical acid activation · X-ray diffractometry · Interplanar distance · Transmission Mössbauer spectroscopy

## 1 Introduction

Clay minerals are aluminosilicates with laminar structure composed of tetrahedral and octahedral sheets situated in special arrangements depending on the type of clay mineral; one of these are the smectites having two tetrahedral and one octahedral sheets per layer.

Y. Huaypar (✉) · J. Bravo
Facultad de Ciencias Físicas, Universidad Nacional Mayor de San Marcos,
Av. Venezuela Cdra. 34 s/n, Lima 01, Perú
e-mail: yhuaypar@yahoo.es

J. Bravo
e-mail: jbravoc@unmsm.edu.pe

A. Gutarra · E. Gabriel
Facultad de Ciencias, Universidad Nacional de Ingenieria, Av. Tupac Amaru 210, Lima 25, Perú

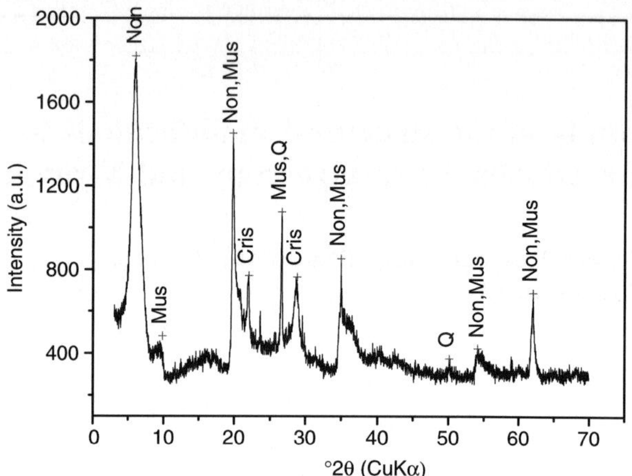

**Fig. 1** Diffractogram of CS01 without acid activation



Their physical properties and chemical compositions are determined by the geological conditions in which they were formed [1].

The smectites are phyllosilicates that are commonly used as non metallic adsorbent materials due to their high adsorption capacity, being used in many industrial processes. In order to increase the adsorption capacity of clays many chemical methods are applied, among these is the acid activation treatment that consists in interchanging a desired proportion of the cations from the octahedral sheets with hydrogen [2].

The adsorption capacity of the clay depends on its mineral composition, cation content and structure. This work tries to study a clay mineral with high iron content in order to monitor the effects caused by the chemical acid activation on the valence states of the iron cations that occupy the structural sites in the clay. TMS is very useful for this purpose.

## 1.1 Experimental

## 1.2 Sample selection and acid activation treatment

The sample used for this work was selected due to its high clay and iron content. The sample was labeled as CS01 and it is a standard commercial smectite. For the respective analyses it was grounded and sieved in a 200 mesh.

For the acid activation treatment, HCl was used in a Pyrex reactor, with reflux condenser and mechanical agitation. The treatment conditions were: solid/acid ratio 6.6% *w/v*, 2 h of treatment time at 90°C, and 2, 4 and 8 N acid concentrations. After the treatment the samples were washed up to pH 3.

## 1.3 Measuring conditions for MS and XRD

$^{57}$Fe transmission Mössbauer spectroscopy, based on the $\gamma$14,41 KeV transition, was used to characterize the Fe mineralogy in the clay. A 25 mCi $^{57}$Co radioactive source in a Rh matrix was used. The Mössbauer spectra were obtained with 250 mg aliquots in a sample holder with an inner diameter of 1.70 cm; all the spectra were taken at room temperature and fitted using the Normos program in its site version [3].

 Springer

**Table 1** Elemental composition for untreated and treated 8 N samples by atomic absorption

| Element (ppm)<br>Sample | Na | K | Mg | Ca | Fe | Al |
|---|---|---|---|---|---|---|
| CS01 | 105,669.8 | 5,953.7 | 14,086.1 | 0 | 21,519.2 | 96,830.7 |
| CS01: 8 N, 2 h, 3 pH | 5,134.8 | 4,501.5 | 11,689.9 | 0 | 18,008.5 | 94,854.3 |

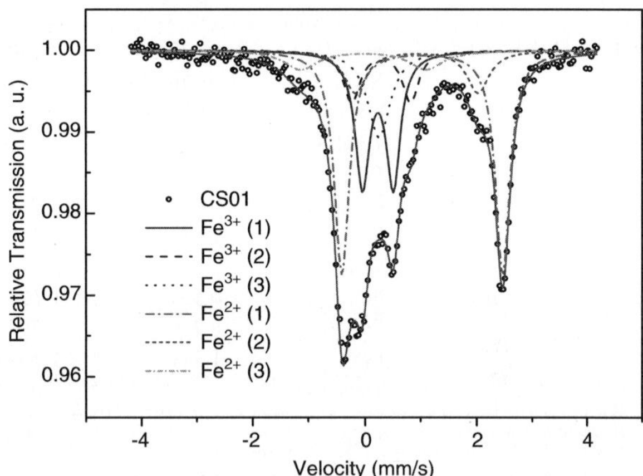


**Fig. 2** Mössbauer spectrum of CS01 without acid activation

For X-ray diffractometry (XRD) analysis a RIGAKU X-ray diffractometer, Miniflex model, was used with Cu-K$\alpha$ radiation and a Ni filter. The measurements were made with $5° < 2\theta < 70°$, $0.02°$ angular step and 2 s/step of counting time. About 500 mg of sample was used using a powder mount.

## 2 Results and discussion

The results of the XRD analysis shows that CS01 is composed mainly of nontronite, as the major component, and of muscovite, as a minor component, that belong to the smectite and mica groups respectively. The position of the characteristic peak corresponding to the inter planar $d_{001}$ distance lies on 14.84 Å for nontronite and 9.97 Å for muscovite. Other components are quartz and cristobalite. See Fig. 1.

For reference, the elemental composition determination of the untreated and treated 8 N samples was made by atomic absorption analysis. The results are shown in Table 1.

The results of the TMS measurements are exhibited in Fig. 2. As starting point, the fitting of this spectrum was carried out using a model that includes two $Fe^{3+}$ doublets to take into account the cis- and trans-sites in nontronite, considering that nontronite may contain almost all of the iron in the sample [4, 5]. Additionally, three $Fe^{2+}$ doublets, including the very strong doublet visible in the spectrum, and a singlet (probably iron in the tetrahedral or interlayer sites [6, 7]), were needed to complete the fitting model; in the last step of this procedure all the parameters were left free to vary. Table 2 gives the hyperfine parameters as well as the partial relative resonance absorption areas for each of the Fe sites present in this sample. The IS value obtained for the doublet $Fe^{2+}$ (3) is rather unusual and requires further study. The IS values are given relative to $\alpha$-Fe.

**Table 2** Hyperfine parameters for CS01 with and without acid activation

| Sample | Site | IS (mm/s) | QS (mm/s) | Area (mm/s) | Area (%) |
|---|---|---|---|---|---|
| CS01 | $Fe^{3+}$ (1) | 0.342 | 0.550 | 0.0156 | 21.94 |
| | $Fe^{3+}$ (2) | 0.405 | 1.050 | 0.0070 | 9.85 |
| | $Fe^{3+}$ (3) | 0.363 | 0.000 | 0.0082 | 11.52 |
| | $Fe^{2+}$ (1) | 1.139 | 2.870 | 0.0271 | 38.12 |
| | $Fe^{2+}$ (2) | 1.039 | 2.193 | 0.0078 | 10.97 |
| | $Fe^{2+}$ (3) | 0.095 | 2.230 | 0.0054 | 7.60 |
| | Total Area | | | 0.0711 | 100.00 |
| CS01: 2 N, 2 h, pH 3 | $Fe^{3+}$ (1) | 0.329 | 0.543 | 0.0414 | 68.35 |
| | $Fe^{3+}$ (2) | 0.376 | 1.374 | 0.0131 | 21.63 |
| | $Fe^{2+}$ | 1.297 | 2.491 | 0.0061 | 10.02 |
| | Total Area | | | 0.0606 | 100.00 |
| CS01: 4 N, 2 h, pH 3 | $Fe^{3+}$ (1) | 0.334 | 0.539 | 0.0447 | 68.71 |
| | $Fe^{3+}$ (2) | 0.417 | 1.263 | 0.0178 | 27.42 |
| | $Fe^{2+}$ | 1.281 | 2.460 | 0.0025 | 3.87 |
| | Total Area | | | 0.0650 | 100.00 |
| CS01: 8 N, 2 h, pH 3 | $Fe^{3+}$ (1) | 0.334 | 0.522 | 0.0385 | 64.14 |
| | $Fe^{3+}$ (2) | 0.390 | 1.294 | 0.0185 | 31.03 |
| | $Fe^{2+}$ | 1.314 | 2.534 | 0.0029 | 4.83 |
| | Total Area | | | 0.0599 | 100.00 |

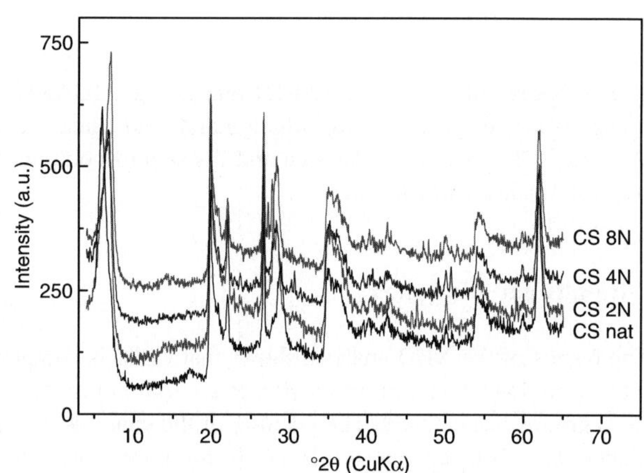


**Fig. 3** Diffractograms of CS01 without and with acid activation at 2, 4 and 8 N

## 2.1 Structural changes in clays due to activation

In order to establish a qualitative description of the structural changes in the clays produced by acid activation at 2, 4 and 8 N, a new sequence of characterization measurements by the above mentioned techniques was undertaken. By XRD a shift in the angular position of the (001) reflection peak was observed (Fig. 3). A slight decrease is observed in the inter planar distance as the position of the peak moves from about 14.7 to 12.6 Å as the acid concentration increases. Moreover, the intensities of the peaks tend to decrease after activation; this effect would indicate a slight dissolution of its structure [8, 9]. There is no presence of muscovite in the sample after acid treatment.

**Fig. 4** Mössbauer spectra of CS01 with acid activation

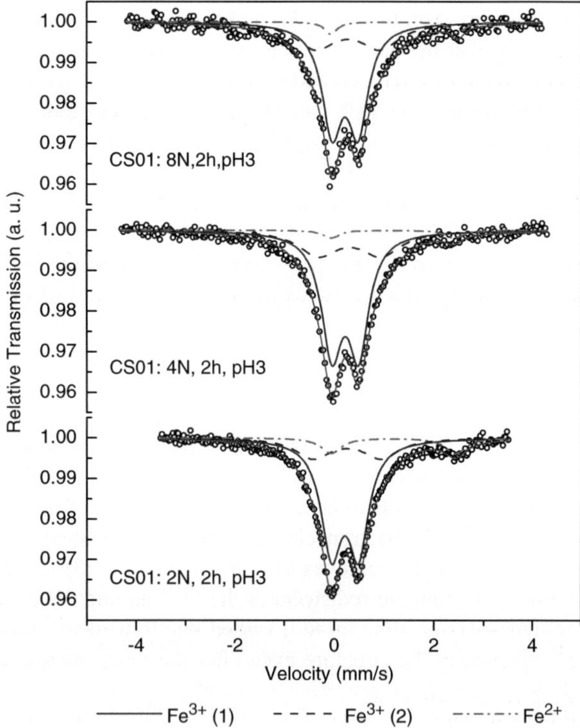


**Fig. 5** Representative curves of relative resonance absorption areas per Fe site versus acid concentration treatment

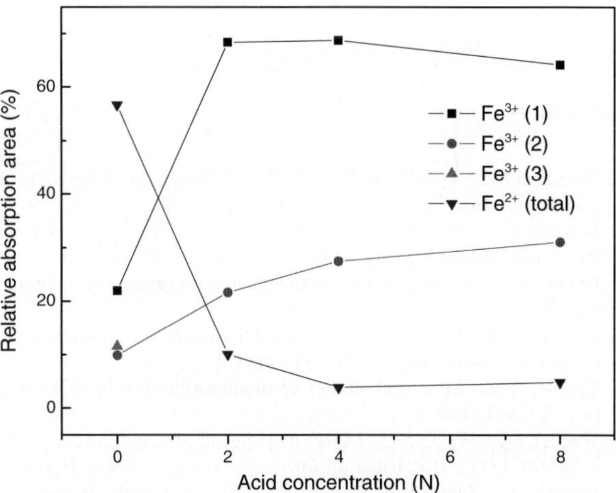


Figure 4 shows the Mössbauer spectra of the acid activated clay at different concentrations. It can be noticed that the total resonance absorption area decreases by about 9% on the average with the acid treatment. For comparison, it is mentioned that in the case of treated at 8 N sample a measurement of iron content was carried out by atomic absorption; the result was a loss of 16.3% of iron which compares with the reduction of 15.8% in the resonance absorption area. With the help of Table 2 the results for the untreated clay with those for the acid treated clay can be compared. This information allows us to infer that the acid treatment at 2 N oxidizes almost all of the $Fe^{2+}$ cations with these staying at the original cis- and trans-sites they were occupying; the acid treatments at 4 and 8 N redistribute the $Fe^{3+}$ cations among cis- and trans-sites with no preference; a small amount of $Fe^{2+}$ remains probably in the octahedral sites. Moreover, the acid treatment removed the $Fe^{3+}(3)$ and part of the $Fe^{2+}(2)$. The QS for $Fe^{3+}(2)$ at trans-sites increased on the average by about 25% due to the acid treatment.

For each of the Fe sites, Fig. 5 shows the relative resonance absorption areas versus the acid concentration. In this way we have an idea of the variations in the relative occupancy by Fe in its two valence states in the structural sites of the clay with the acid treatment.

## 3 Conclusions

According to the XRD results the acid activation affects only the (001) reflection peak of the nontronite, which shows a decrease of its interplanar distance and intensity. The Mössbauer analyses help us to describe the effects of the acid treatment on the clay as view with reference to the iron cations. The acid treatment oxidizes most of the $Fe^{2+}$ in the octahedral sites and removes the iron at the $Fe^{3+}$ (3) site and some of the $Fe^{2+}$. The more intense acid treatment redistributes the $Fe^{3+}$ among the octahedral sites with no preference. The trans-sites become more distorted due to acid treatment. And a very small amount of $Fe^{2+}$ remains in the structure even after the more intense acid treatment.

**Acknowledgments** The authors thank the X-ray Diffractometry and Mössbauer Spectroscopy Laboratories at the Facultad de Ciencias Físicas, Universidad Nacional Mayor de San Marcos, and the Catalysis Laboratory at Universidad Nacional de Ingenieria for their services.

## References


1. Moore, D.M., Reynolds Jr, R.C.: X-Ray Diffraction and the Identification and Analysis of Clay Materials, 2nd edn. Oxford University Press, Oxford (1997)
2. Volzone, C., Porto Lopez, J.M., Pereira, E.: Acid activation on smectitic material: I. structural analysis. Rev. Latinoamericana. Ing. Quím. Apl. **16**, 205–215 (1986)
3. Brand, R.A.: Normos Mössbauer Fitting Program, User's Guide. Wissenschaftlich Elektronik GmbH, Starnberg (1995)
4. Cardile, C.M., Johnston, J.H.: 57 Fe Mössbauer spectroscopy of montmorillonites: a new interpretation. Clays Clay Miner **34**, (3), 307–313 (1986)
5. Cardile, C.M.: Structural studies of montmorillonites by 57Fe Mössbauer spectroscopy. Clay Miner **22**, (4), 387–394 (1987)
6. Wagner, F.E., Wagner, U.: Mössbauer spectra of clays and ceramics. Hyperfine Interact **154**, 35–82 (2004)
7. Rancourt, D.G.: Mössbauer spectroscopy in clay science. Hyperfine Interact **117**, 3–38 (1998)
8. Volzone, C., Zalba, P.F., Pereira, E.: Activación ácida de esmectitas. II estudio mineralógico. An. Soc. Quím. Argent. **76**, (1), 57–68 (1988)
9. Stucki, J.W., Lee, K., Zhang, L., Larson, R.A.: Effects of iron oxidation state on the surface and structural properties of smectites. Pure Appl. Chem. **74**, (11), 2145–2158 (2002)

Hyperfine Interact (2007) 175:29–34
DOI 10.1007/s10751-008-9584-1

# Magnetic and structural properties of mechanically alloyed $Tb_{0.257-x}Nd_xFe_{0.743}$ alloys, with $x=0$ and $0.257$

Y. Rojas Martinez · H. Bustos Rodriguez ·
D. Oyola Lozano · G. A. Pérez Alcázar · J. C. Paz

Published online: 1 April 2008
© Springer Science + Business Media B.V. 2008

**Abstract** The alloys between a transition metal and a rare earth present magnetic and magneto optical properties of exceptional interest for the production of magnetic devices for information storage. In this work we report the magnetic and structural properties, obtained by Mössbauer spectrometry (MS) and X-ray diffraction (XRD), of $Tb_{0.257-x}Nd_xFe_{0.743}$ alloys with $x=0$ and $0.257$ prepared by mechanical alloying during 12, 24 and 48 h, to study the influence of the milling time in their magnetic and structural properties. The X-rays results show for all the samples that the $\alpha$-Fe and an amorphous phase are always present. The first decreases and the second increases with the increase of the milling time. Mössbauer results show that the amorphous phase in samples with Nd is ferromagnetic and appears as a hyperfine field distribution and a broad doublet, and that as the milling time increases the paramagnetic contribution increases. For samples with Tb the amorphous phase is paramagnetic and appears as a broad doublet which increases with the milling time and for 48 h milling it appears an additional broad singlet.

**Keywords** Mössbauer spectrometry · Mechanical alloying · X-rays · Magneto-optical

## 1 Introduction

The rare-earth-transition metal (RE-TM) amorphous alloys have attracted considerable attention as a result of their applications in magneto optical recording. The basic magnetic and magneto-optical properties of these alloys can be well tailored via selection of appropriate preparation conditions and compositions. One of the required conditions for high density magneto-optic recording media is the presence of perpendicular magnetic anisotropy (PMA). In 1956, Meiklejohn and Bean [1, 2] reported how this anisotropy is the

Y. R. Martinez (✉) · H. B. Rodriguez · D. O. Lozano
Department of Physics, University of Tolima, A.A. 546, Ibagué, Colombia
e-mail: yarojas@ut.edu.co

G. A. P. Alcázar · J. C. Paz
Department of Physics, University of Valle, A.A. 25360, Cali, Colombia

result of the interaction between an antiferromagnetic and a ferromagnetic material. The exchange anisotropy phenomenon has become the basis for important applications in information storage technology. The perpendicular anisotropy of amorphous rare-earth-transition metal (RE-TM) thin films has been a challenge since it was first observed in 1973 [3]. It has been known that Tb-Fe based alloys have a strong perpendicular magnetic anisotropy, and they have been used as magneto-optical (MO) material for disk [4]. In general, the magnetic properties of RE-TM amorphous alloys depend on preparation conditions. The origin of PMA in RE-TM thin-film materials and RE-TM alloys is not yet fully understood. A variety of mechanisms have been proposed to explain the PMA [3–5], ranging from internal stresses to columnar microstructures, but the existence of a growth-induced structural anisotropy seems to be strongly correlated with the magnetic anisotropy. Hansen et al. [6] investigated the magnetization, Curie temperature, uniaxial anisotropy, coercivity, and Faraday and Kerr rotations as a function of composition and temperature for amorphous RE-TM alloys of compositions $RE_{1-x}Fe_x$ with RE=Nd, Tb, Pr with $0<x<1$, and intended to provide a new generation of magneto optical disk for recording. Dancygier [7] studied the $Tb_{0.257}Fe_{0.743}$ thin films system and reported the magnetic properties of this prepared composition. The demand for these novel materials, whit improved properties, in this technologically advancing word, has made powder metallurgy a very important field of materials research. Inside this field mechanical alloying is one of the most efficient methods to obtain nanostructure materials which can be used for the manufacture of targets in order to growth thin films with magneto optical properties.

In order to study the influence of the processing time and concentration on the structural and magnetic properties of the $Tb_{0.257-x}Nd_xFe_{0.743}$ alloys system, with $x=0$ and 0.257, they were prepared by mechanical alloying during 12, 24, 36 and 48 h and characterized by XRD and MS.

## 2 Experimental procedure

Iron–terbium and iron–neodymium powders were mixed in the compositions $Tb_{0.257}Fe_{0.743}$ and $Nd_{0.257}Fe_{0.743}$ and then mechanically processed in argon atmosphere during 12, 24, 36 and 48 h. A Fritsch pulverisette 7 high energy planetary ball mill, with hardened stainless steel vials of 50 ml of volume and balls of the same material with 11 mm of diameter, was used. A speed of 280 rpm and a ball to powder mass ratio of 20:1 were used. Neodymium powder with an average particle size of 40 meshes and terbium chips of 0.5 mm were employed. Mössbauer spectra were obtained at room temperature in transmission geometry using a conventional constant acceleration spectrometer with a 57Co-Rh source. The spectra were fitted with sextets and doublets using the MOSFIT program [8]. The $\alpha$-Fe pattern was used as calibration sample. The X-ray analysis to establish the structure of the lattice were performed at room temperature for all samples using a RINT20001 diffractometer with the Cu K$\alpha$ radiation and the patterns were refined by using Maud program [9].

## 3 Results and discussion

Figure 1 shows the XRD patterns of the $Tb_{0.257-x}Nd_xFe_{0.743}$ powders with $x=0$ and $x=0.257$ milled during 12, 24 and 48 h. It can be seen, in all the patterns, the peaks of $\alpha$-Fe and an amorphous phase. Then all the RE atoms are alloyed with Fe in an amorphous

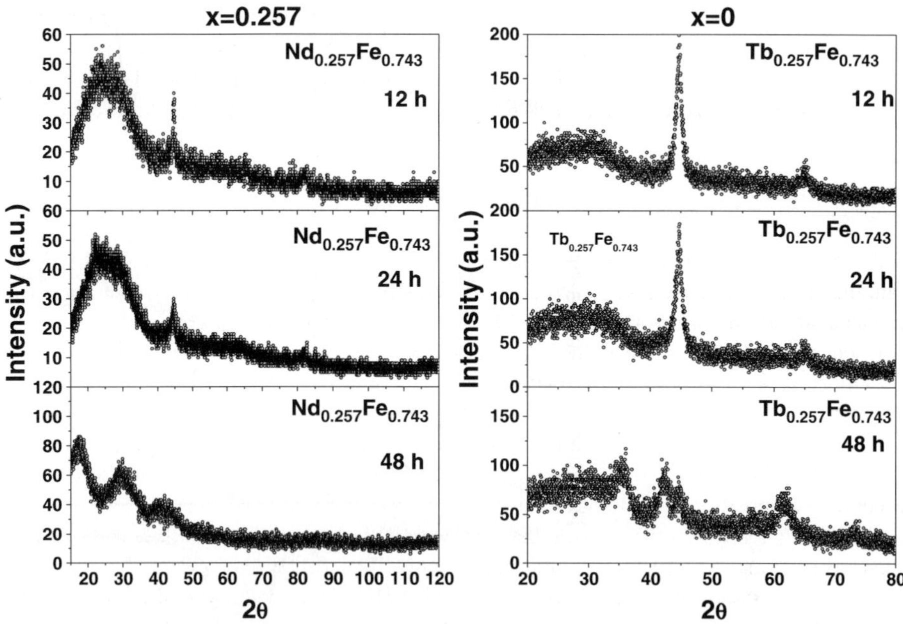


**Fig. 1** X-ray diffraction pattern of Tb$_{0.257-x}$Nd$_x$ Fe$_{0.743}$ powders with $x=0$ and $x=0.257$

phase. It is clear from this figure that when the milling time increases, the intensity of the $\alpha$-Fe peaks decreases while that the amorphous phase increases. It can be noted in this figure that alloys with Nd present a bigger quantity of amorphous phase, then lest intensive $\alpha$-Fe peaks. No one peak of RE elements or of oxides were detected showing that they are not present or their volume quantity is lest than 2%. The XRD patterns of 48 h show very broad and weak Fe lines showing that it is present as nano grain or nearly amorphous way.

Figure 2 shows the MS obtained at room temperature of the Tb$_{0.257-x}$Nd$_x$Fe$_{0.743}$ powders with $x=0.257$ milled during 12, 24 and 48 h. For 12 and 24 h the spectra were fitted by means of a sextet, a hyperfine field distribution (HFD), and a broad doublet. The sextet is associated to $\alpha$-Fe, and the HFD and the broad doublet are associated with an iron-based amorphous phase which presents Fe sites rich in Fe nearest neighbors (the HFD) and poor in Fe nearest neighbors (broad doublet), respectively. The obtained mean hyperfine fields of the HFDs were 24.2 T and 30.5 T for 12 and 24 h, respectively. For the 48 h sample the fit were conducted with the sextet of the $\alpha$-Fe and the broad doublet, then the amorphous phase is now originated by the Fe paramagnetic and asymmetric sites. It can be observed that the sextet associated to $\alpha$-Fe decreases with milling time, while the doublet associated to the amorphous iron-based phase increases. Then the milling produces an amorphous alloy of Fe and Nd in such way that for 12 h it is ferromagnetic but when the milling continue it turn to a paramagnetic behaviour. Table 1 shows the Mössbauer parameters obtained from the fit.

Figure 3 shows the MS obtained at room temperature of the Tb$_{0.257-x}$Nd$_x$Fe$_{0.743}$ powders with $x=0$ milled during 12, 24 and 48 h. The spectra were fitted with two components: a sextet associated to the $\alpha$-Fe, and a broad paramagnetic site (doublet) associated to the amorphous Fe based phase. The spectrum for 48 h presents an additional component fitted with a broad paramagnetic single line showing that for these conditions the spectrum presents symmetric (singlet) and asymmetric paramagnetic (doublet) sites. In according

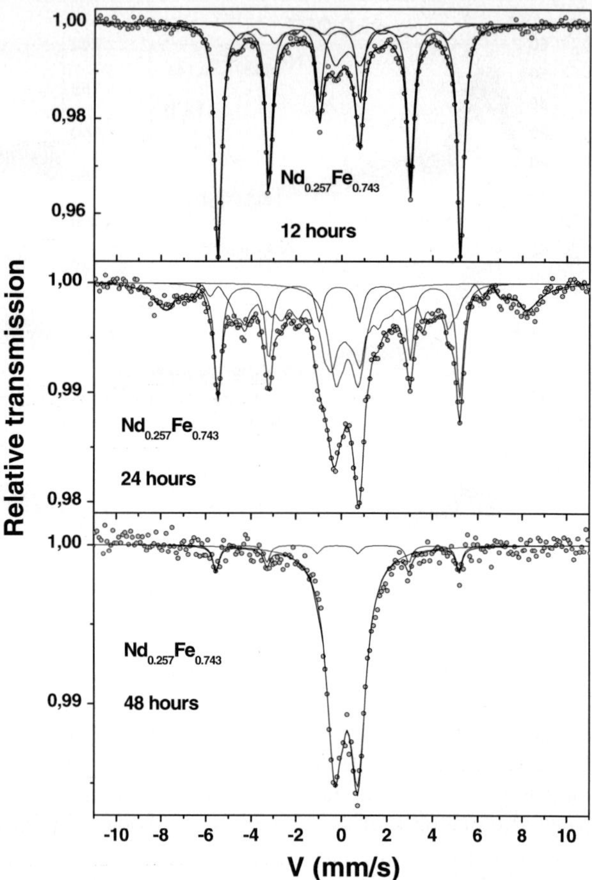



**Fig. 2** Mössbauer spectra of Nd$_{0.257}$Fe$_{0.743}$ samples prepared by MA at different times

**Table 1** Mössbauer parameters of Tb$_{0.257-x}$Nd$_x$Fe$_{0.743}$ powders with $x=0.257$

| Nd$_{0.257}$Fe$_{0.743}$ | $\delta$ | $\Gamma$ | SQ | MHF | Percent |
| --- | --- | --- | --- | --- | --- |
| 12 h | | | | | |
| Doublet | 0.39 | 0.76 | −0.98 | | 12 |
| α-Fe | 0.04 | 0.35 | −0.01 | 32.9 | 73 |
| HFD | 0.09 | | −0.12 | 24.2 | 15 |
| 24 h | | | | | |
| Doublet | 0.41 | 0.76 | 0.89 | | 20 |
| α-Fe | 0.06 | 0.37 | −0.04 | 32.9 | 25 |
| HFD | 0.26 | | 0.25 | 30.5 | 55 |
| 48 h | | | | | |
| Doublet | 0.39 | 0.83 | 0.91 | | 91 |
| α-Fe | 0.00 | 0.33 | −0.02 | 33.0 | 9 |

Mean hyperfine field (MHF) values are in Tesla, the isomer shift ($\delta$), line width ($\Gamma$) and the quadrupolar splitting (SQ) are in mm/s

 Springer

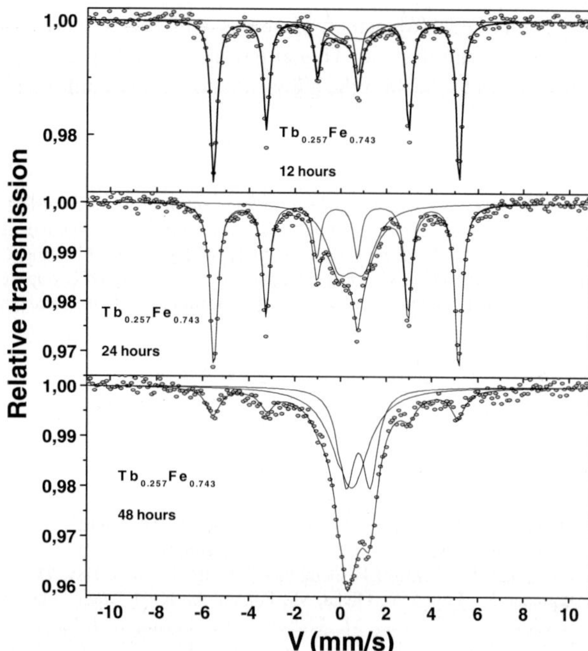

**Fig. 3** Mössbauer spectra of Tb$_{0.257}$Fe$_{0.743}$ samples prepared by mechanical alloying



**Table 2** Mössbauer Parameters of Tb$_{0.257-x}$Nd$_x$ Fe$_{0.743}$ powders with $x=0$

| Tb$_{0.257}$Fe$_{0.743}$ | $\delta$ | $\Gamma$ | SQ | MHF | Percent |
|---|---|---|---|---|---|
| **12 h** | | | | | |
| Doublet | 0.67 | 1.26 | 0.96 | | 14 |
| $\alpha$-Fe | 0.00 | 0.32 | −0.04 | 33.0 | 86 |
| **24 h** | | | | | |
| Doublet | 0.63 | 1.26 | 0.98 | | 36 |
| $\alpha$-Fe | 0.00 | 0.38 | −0.02 | 32.9 | 64 |
| **48 h** | | | | | |
| Singlet | 0.64 | 1.90 | | | 55 |
| Doublet | 0.92 | 0.72 | 1.0 | | 33 |
| $\alpha$-Fe | 0.00 | 0.33 | −0.08 | 32.9 | 12 |

Mean hyperfine field (MHF) values are in Tesla, the isomer shift ($\delta$), line width ($\Gamma$) and the quadrupolar splitting (SQ) are in mm/s

with the XRD results the broad doublet can be attributed to the amorphous phase which is paramagnetic, and the sextet to the $\alpha$-Fe, respectively. Then the alloying process produces an amorphous alloy between Fe and Tb which is paramagnetic and its fraction increases with the milling time. In Table 2 the obtained hyperfine parameters are shown.

## 4 Conclusions

Mechanical alloying has been used to prepare samples of the Tb$_{0.257-x}$Nd$_x$Fe$_{0.743}$ system, with $x=0$ and $x=0.257$. All the samples present an amorphous phase, characteristic of RE-

TM compounds [6], which increases with the milling time and present a paramagnetic behavior in samples with Tb and ferromagnetic but tending to be paramagnetic for samples with Nd. Up to the 48 h the alloys are not totally conformed and present a small quantity of Fe.

**Acknowledgements** We are grateful to the Central Committee of Research of the University of Tolima for its financial support, to the Mössbauer Laboratory of the University of Valle, and the support of the Excellence Center for Novel Materials (CENM), for the achievement of this work. We are grateful to the Laboratory Materials of Cinvestav, Mexico and laboratory of Group Science of Materials (Dr. Jairo Roa) of the Universidad Nacional, Colombia, for performing XRD measurements.

# References

1. Meiklejohn, W.H., Bean, C.P.: Phys. Rev. **102**, 141 (1956)
2. Meiklejohn, W.H., Bean, C.P.: Phys. Rev. **105**, 904 (1957)
3. Chaudhari, P., Cuono, J.J., Gambino, R.J.: IBM J. Res. Dev. **11**, 66 (1973)
4. Yamada, K., Fujita, A., Ohata, M., Fukamichi, K.: J. Magn. Mag. Mater. **239**, 412 (2002)
5. Katayama, T., Miyasaki, M., Nishihara, Y., Shibara, T.: J. Magn. Magn. Mater. **35**, 235 (1983)
6. Hansen, P., Raasch, D., Mergel, D.: J. Appl. Phys. **75**, 5267 (1994)
7. Dancygier, M.: IEEE.Trans. Magn. **23**, 2608 (1987)
8. Teillet, J., Varret, F.: Université du Maine, unpublished Mosfit program
9. Lutterotti, L., Scardi, P.J.: Appl. Crystallogr. **23**, 246 (1990)

Hyperfine Interact (2007) 175:35–41
DOI 10.1007/s10751-008-9585-0

# Mössbauer and magnetic studies of parent material from argentine pampas soils

J. C. Bidegain · A. A. Bartel · F. R. Sives ·
R. C. Mercader

Published online: 28 March 2008


**Abstract** In order to establish a correlation between the different types of soils using hyperfine and magnetic parameters as climatic and environmental proxies, we have studied the differentiation of soil developed around 38.5° south latitude, in the central Pampas of Argentina, by means of Mössbauer spectroscopy and environmental magnetism. The soils transect (climosequence) investigated stretches from the drier west (around 64° W) to the more humid east (at around 59° W) in the Buenos Aires Province, covering a distance of 600 km. The soils studied developed during recent Holocene geologic times in a landscape characterized by small relict plateaus, slopes and depressions, dunes and prairies. The parent material consists of eolian sandy silts overlying calcrete layers. The low mean annual precipitation in the western parts of the region gives rise to soils without B-horizons, which limits the agricultural use of land. The preliminary results show an increase of the paramagnetic $Fe^{3+}$ relative concentration from west to east in the soils investigated. Magnetite is probably mainly responsible for the observed enhancement in the susceptibility values. The magnetic response of the parent material is similar to that of the loess part of the previously investigated loess–paleosol sequences of the Argentine loess plateau.

**Keywords** Argentine soils · Magnetism · Paleoclimate · Mössbauer spectroscopy · Parent materials

J. C. Bidegain
Laboratorio de Entrenamiento Multidisciplinario para la Investigación Tecnológica, Calle 52 e/ 121 y 122, 1900 La Plata, Argentina

A. A. Bartel
Facultad de Ciencias Exactas y Naturales, Universidad Nacional de La Pampa,, Av. Uruguay 151, 6300 Santa Rosa, Argentina

F. R. Sives · R. C. Mercader (✉)
Departamento de Física, Facultad de Ciencias Exactas, Universidad Nacional de La Plata, IFLP, CC 67, 1900 La Plata, Argentina
e-mail: mercader@fisica.unlp.edu.ar

## 1 Introduction

Several studies concerned with the Argentine loess Plateau indicate that the magnetic susceptibility values and the enhancement of intensity are due to the higher amount of titanomagnetites in the parent materials [1]. All the Quaternary sections studied up to now indicate that the specific susceptibility values increase in loess and decrease in soils [2, 3]. To understand the relation between magnetic and paleoclimatic records, some time ago, some of us set about studying the magnetic and hyperfine properties of loess–paleosol sequences in loess deposits of Argentina [4, 5].

Bidegain et al. [6] mention the existence of a third magneto-climatic model for the Argentine loess deposits that differ from the wind vigor model (prevailing in Siberian soils) and the pedogenetic model (widespread in the Chinese sequences). In the Argentine case, the western winds during glacial intervals entrain large quantities of dense magnetic particles than during interglacial periods. However, pedogenesis and weathering cause the depletion of magnetic signals by alteration of titanomagnetites and the formation of new paramagnetic and antiferromagnetic iron oxides (particularly hematite) and oxyhydroxides (goethite, lepidocrocite, akaganeite). Pedogenesis does not affect the original material in the same way along the sedimentary sequences and consequently the behavior of magnetic parameters is more complex.

Because of the high-economic significance of the Pampean soils from an agricultural point of view, we are investigating the influence of the pluvial regimes on the model of magnetic records across an east–west transect of about 600 km by measuring the same magnetic parameters as those of [6]. According to the measurements of rock magnetic parameters carried out in soils horizons, the mode of magnetic pattern around 38° S indicates that susceptibility values increase upwards in the soils profiles. The opposite has been mentioned in La Plata area in [6], giving rise to the idea that the rainfall may control the magnetic signal.

We are trying to relate the current weather conditions to the older ones affecting particularly the uppermost layers of the soil profiles. To this end, we find it necessary to evaluate if the higher magnetic parameters of the uppermost A- and B-horizons have some bearing on those of the parent materials. Since the parent material is a key factor in the soil-forming process, its characterization through magnetic and hyperfine parameters is important to assess the lithogenetic and pedogenetic contributions. In this work we report the preliminary results of the investigations that we have performed so far on parent material samples.

## 2 Experimental

The soil samples analyzed in this work correspond to the parent material of six representative profiles carried out along a transect across the ≈38.5° parallel to the south of the Ventania range from ≈64° W (arid soils) to ≈59° W (udic soils). Grain size determination, magnetic concentration and mineralogical analysis are been performed aiming at establishing a correlation with the Mössbauer data reported in the present contribution.

The parent materials are sandy to silty loess (Table 1) lying discordantly on a "calcrete layer" attributed to a drier climate stage during Plio-Pleistocene. The petrocalcic horizon is at a very shallow depth at the base level of all the soil-profiles indicated in Fig. 1 and marks a hiatus representative of one or more denudation periods, during which the overlying

 Springer

**Table 1** Characterization of the studied soils according to the USA soil taxonomy

| Soils | Moisture regime | Soil taxonomy | Texture | pH | Color (Munsell moist) | $CaCO_3$ (%) |
|---|---|---|---|---|---|---|
| 1 | Aridic | Haplocalcid | Sandy | 7.4 | 10 YR 4/3 | 4.5 |
| 2 | Ustic | Typical Haplustol | Sandy loam | 7.52 | 10 YR 5/3 | 14.6 |
| 3 | Ustic | Typical Argiustol | Sandy loam | 7.06 | 10 YR 4/3 | s/r |
| 4 | Ustic | Haplustol | Sandy | 6.35 | 10 YR 4/3 | s/r |
| 5 | Udic | Hapludol | Sandy | 6.77 | 10 YR 4/3 | s/r |
| 6 | Udic | Argiudol | Sandy loam/loam | 6.57 | 7.5 YR 5/4 | s/r |

Texture, color, pH, and calcium carbonate for the samples are indicated.

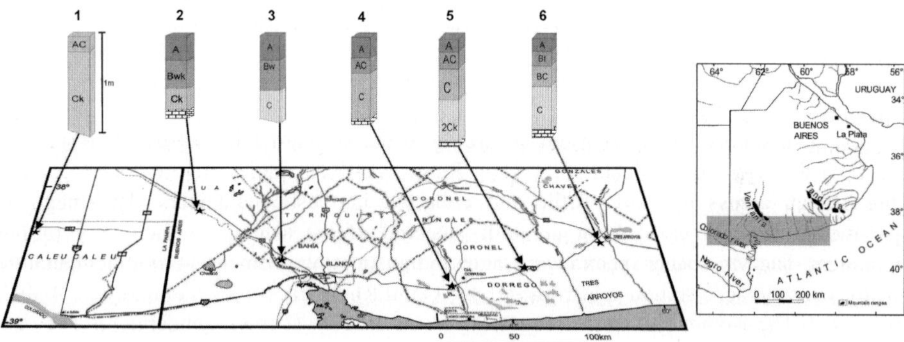


**Fig. 1** Regional map showing the geographical location of the studied area. The labels indicate the soil profiles where samples 1 to 6 were taken

sediments would have been eroded [8]. In this area it is hard to dig profiles deeper than about 50 cm, particularly when the soils are at the top of small calcrete hills. Conversely, thicker soil profiles develop in valleys in-between those small hills that can reach a depth of more than 100 cm.

The geologic age of the studied soils is estimated as belonging to the Holocene (<10 ka). Quaternary sediments (Pleistocene–Holocene) lie discordantly on continental Pliocene sediments in the studied area (Fig. 1). The Pleistocene sediments have a thickness of 30 to 40 m in some places and more than 200 m in others. According to the geological survey, there is no influence of the basement on the mineralogical composition of the soils studied.

The source of the Quaternary eolian sediments seems to be a huge deflation area placed to the northwestern of Patagonia and the central Andes. The general mineralogical composition observed in the area is homogeneous with a clear predominance of light over heavy minerals. Among the former, prevail lithic fragments, quartz, and volcanic glass that generally is altered. Sodium feldspar and plagioclase are present but in less proportion than quartz. Among heavy minerals, pyroxene, magnetite, amphibole and lithic fragments are found [9].

The mineralogical study indicates that the parent material contains volcanic glass in all samples, the amount of sharp-edged fragments of bubble rims (volcanic shards) increases noteworthy towards the west.

The parent material of soil-profiles labeled 4 and 5 exhibit a larger contribution of the sand fraction than profiles 2, 3 and 6. Such textural difference can also be related to the lack of the illuvial subsurface horizon Bt.

**Table 2** Mössbauer hyperfine parameters obtained from spectra at room temperature following the procedure described in the text

| | Sextet | | | | $Fe^{2+}$ doublet | | | $Fe^{3+}$ doublet | | |
|---|---|---|---|---|---|---|---|---|---|---|
| | H (kOe) | ε (mm/s) | δ (mm/s) | RA (%) | δ (mm/s) | Δ (mm/s) | RA (%) | δ (mm/s) | Δ (mm/s) | RA (%) |
| 1 | 497 (4) | −0.08 (6) | 0.27 (6) | 23 (6) | 1.13 (12) | 2.28 (24) | 35 (4) | 0.34 (9) | 0.60 (17) | 42 (3) |
| 2 | 493 (5) | −0.05 (7) | 0.39 (7) | 37 (7) | 1.16 (4) | 2.21 (8) | 19 (5) | 0.33 (1) | 0.57 (7) | 44 (4) |
| 3 | 498 (6) | −0.08 (8) | 0.39 (9) | 29 (8) | 1.24 (8) | 2.20 (15) | 7 (3) | 0.33 (2) | 0.62 (3) | 64 (2) |
| 4 | 504 (6) | −0.10 (9) | 0.33 (10) | 34 (12) | 1.33 (10) | 2.2 (2) | 11 (3) | 0.34 (3) | 0.67 (6) | 55 (6) |
| 5 | 490 (8) | −0.20 (10) | 0.38 (11) | 26 (12) | 1.12 (20) | 2.42 (39) | 14 (8) | 0.39 (6) | 0.58 (11) | 60 (8) |
| 6 | 500 (3) | −0.11 (4) | 0.34 (4) | 23 (6) | 1.18 (9) | 2.27 (17) | 9 (3) | 0.35 (1) | 0.58 (2) | 68 (2) |

The numbers between parentheses are the uncertainties in the least significant figures of the values reported. *RA* relative area

The field survey was carried out according to the standard rules set by the Argentine Institute for Agricultural Technology [10]. Representative samples were extracted each 5 and 10 cm across the soil profile and collected in plastic containers. The magnetic parameters were obtained from the bulk material of the samples without any further treatment. The Mössbauer spectra were taken in transmission geometry at room temperature with a constant acceleration spectrometer. A $^{57}Co$ in Rh matrix source nominally of 10 mCi was used. The isomer shifts (δ) are referred to metallic Fe at room temperature. The data were fitted to Lorentzian line-shapes using non-linear least-squares program with constraints. Three hyperfine sites, namely, one magnetic split signal and two quadrupole doublets were necessary to fit the spectra. No restrictions were imposed on the values of the hyperfine field, the isomer shift or the quadrupole shift, which converged to the values shown in Table 2. This procedure yielded consistent results in spite of the fact that the total amount of iron in each sample is a low percentage of the material of the whole sample.

The alternating current (AC) susceptibility measurements between 13 and 325 K were carried out in a LakeShore 7130 AC susceptometer using a field amplitude of 1 Oe, and a frequency of 825 Hz.

## 3 Results and discussion

The magnetic susceptibility shows clearly the typical behavior of that of the Pampas loess deposits. The Verwey transition characteristic of magnetite can be observed analyzing the samples across the transect (see Fig. 2). The slight shift of the transition temperatures might be related to the substitution of Fe for Ti or Al in magnetite. As titanomagnetites have been reported in similar studies [3], we assume that the magnetite found either belongs or is close to that of the end member of titanomagnetite solid-solution series. The presence of an intermediate composition is not apparent in the current magnetic response. We would need to correlate the total Fe content with the mineralogical characteristics of the different samples and horizons. It is worth noting, however, that all the studied samples show the same trend. This response is similar to that obtained in loess deposits reported in previous works [5].

From a magnetic or hyperfine point of view, minerals and soils are complex systems. Applying these techniques to samples belonging to extensive regions, we can set apart the

 Springer

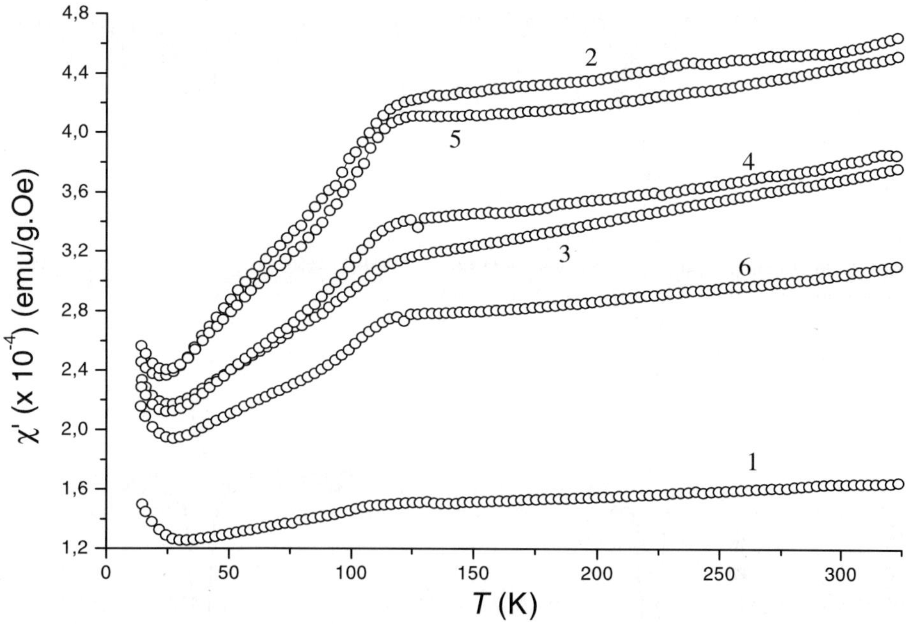

**Fig. 2** Thermal dependence of the in-phase AC magnetic susceptibility for the samples indicated in the Fig. 1

local variation from the more general trends. This strategy is adequate to understanding essential features that otherwise do not allow establishing patterns of behavior. These need to be correlated with the size and the texture of each soil. No specific pattern was found on the magnetic susceptibility or in the magnetic iron contents, even when similar thermal magnetic susceptibility behavior (Fig. 2) was found.

Because of the low iron content of the bulk parent material, the statistics of the Mössbauer spectra (Fig. 3) did not allow fittings assuming more than three kinds of sites, one magnetic and two paramagnetic ones. In addition, because of the broad line widths and poor statistics, the area of the magnetic split signal (Table 2) should be considered as the sum of the contributions arising from all the iron oxides exhibiting magnetically split regimes at room temperature in the sample.The magnetic split signals show small variations on the hyperfine parameters, likely as a result of the different hematite to magnetite–maghemite contents. However, we cannot find a straightforward correlation with the change of the magnetic susceptibility, chiefly because of the existence of a relative significant magnetite concentration with respect to other magnetic minerals in the sample. For all samples, the fittings yielded similar paramagnetic parameters for the $Fe^{2+}$ sites ($\delta \approx 1.3 \pm 0.1$ mm/s and $\Delta \approx 2.2 \pm 0.2$ mm/s) and $Fe^{3+}$ sites ($\delta \approx 0.35 \pm 0.05$ mm/s and $\Delta \approx 0.6 \pm 0.1$ mm/s). This is coherent with the mineralogical information mentioned above.

The behavior of sample 3 does not follow the same trend as the other samples, both from the hyperfine and magnetic point of view.

No obvious correlation is observed between the different values obtained. However, even for these preliminary results, there is a clear trend: the area of paramagnetic $Fe^{3+}$ decreases from east to west (Table 2). One might be tempted to associate this to the small variation on the texture (coarser to the west), connect it with a variation on the relative mineralogical composition, and then, associate it with a wind model. However, a more

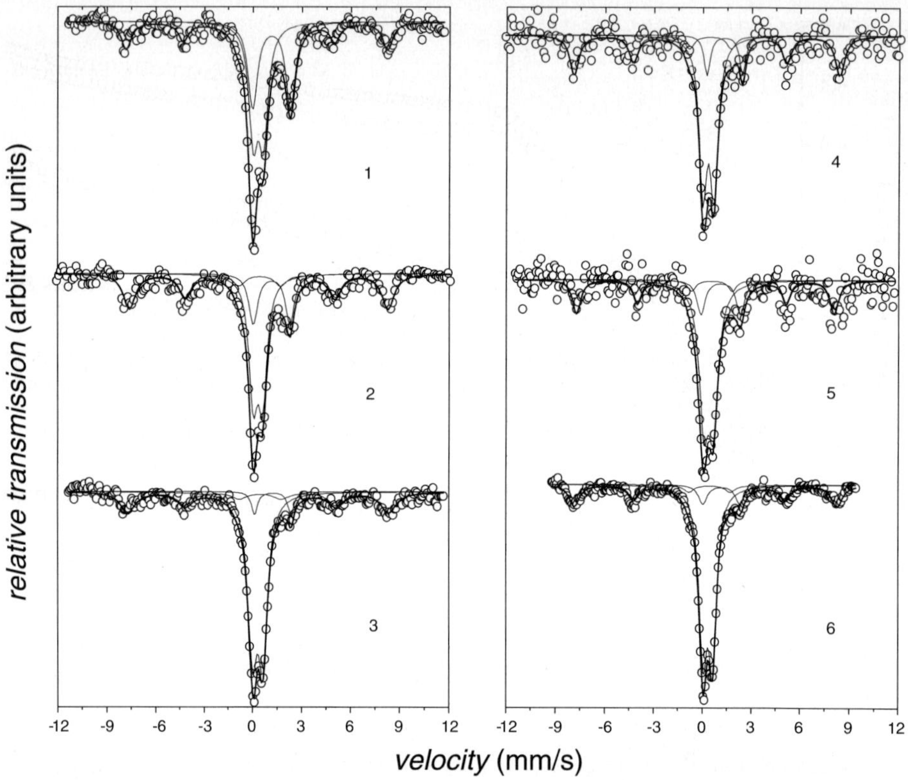

**Fig. 3** Mössbauer spectra of bulk samples. The *open circles* are the experimental points. The *thick solid line* is the least squares fitting to the experimental data. The *thin solid lines* are the components of the three hyperfine sites used in the fitting

detailed mineralogical study would be necessary to proceed in this manner. Current studies about the lithological and mineralogical changes of the soils parent materials along the transect, may help us arrive at unambiguous conclusions when carefully taken into account.

## 4 Conclusions

The magnetic response of the parent material across the transect is similar to that of loess deposits of the northern Pampas. Like those sediments, it has probably an eolian origin from the west.

The current results do not show a direct correlation of the magnetic susceptibility with the Mössbauer relative percentages of the magnetic split areas in the spectra.

There is a west to east increase of Mössbauer paramagnetic $Fe^{3+}$ signals, which, based on the current preliminary results, don't allow establishing a plausible cause for the variation. However, the increase in pedogenetic $Fe^{3+}$ in the most developed soils to the eastern part of the region, which may reflect the increase in the rainfall, should not be discarded. Ongoing magnetic separation and material sieving will help us determine a model of behavior.

**Acknowledgments** The authors thank Consejo Nacional de Investigaciones Científicas y Técnicas (CONICET), Comisión de Investigaciones Cientificas de la Provincia de Buenos Aires (CICPBA), and Agencia Nacional de Promoción Científica y Tecnológica for partial economic support. JCB is member of CICPBA, AAB is fellow of CONICET, and RCM and FRS are members of CONICET, Argentina.

# References


1. Bidegain, J.C., Pavlicevic, R., Iasi, R.R., Pérez, R.H.: Susceptibilidad Magnética y Concentraciones de FeO en Loess y Paleosuelos Cuaternarious como Indicadores de Cambios Paleoambientales y Paleoclimáticos. Proceedings XIII geology and III oil exploration congress. Argentina. **II**, 521 (1996)
2. Bidegain, J.C.: New Evidence of Brunhes/Matuyama Polarity boundary in the Hernández-Gorina Quarries, North-West of de city of La Plata, Buenos Aires Province, Argentina. Quatern. South Am. Antarct. Peninsula. **II**, 207 (1998)
3. Bidegain, J.C., Rico, Y.: Mineralogía magnética y registros de susceptibilidad en sedimentos cuaternarios de polaridad normal (Brunhes) y reversa (Matuyama) de la cantera de Juárez, provincia de Buenos Aires.. Rev Asoc Geol Argent. **59**, 451 (2004)
4. Terminiello, L., Bidegain, J.C., Rico, Y., Mercader, R.C.: Characterization of Argentine loess and paleosols by Mössbauer spectroscopy. Hyperfine Interact. **136**, 97 (2001)
5. Mercader, R.C., Sives, F.R., Imbellone, P.A., Vandenberghe, R.E.: Magnetic and Mössbauer studies of Quaternary Argentine loessic soils and paleosols.. Hyperfine Interact. **161**, 43 (2005)
6. Bidegain, J.C., Evans, M.E., van Velzen, A.J.: A magnetoclimatological investigation of Pampean loess. Geophys. J. Int. **160**, 55 (2005)
7. Bidegain, J.C.: Sedimentary development, magnetostratigraphy and sequence of events of the Late Cenozoic in Entre Ríos and surrounding areas in Argentina. Ph.D. Thesis, University of Stockholm, Stockholm, (1991)
8. González Uriarte, M.: Características geomorfológicas de la porción continental que rodea la Bahía Blanca, Provincia de Buenos Aires. IX Congreso Geológico Argentino, Argentina (1984)
9. Lüters, J.: Edafogénesis de la climosecuencia existente entre el sureste de la provincia de La Pampa y el litoral Atlántico. Magister Thesis, Universidad Nacional del Sur, Argentina (1982)
10. Etchevehere, P.: Normas de reconocimiento de suelos, vol 152, 2nd edn. p. 211. Departamento de Suelos, INTA, Castelar (1976)

Hyperfine Interact (2007) 175:43–48
DOI 10.1007/s10751-008-9586-z

# Multivariate analysis in provenance studies: Cerrillos obsidians case, Peru

A. Bustamante · M. Delgado · R. M. Latini ·
A. V. B. Bellido

Published online: 1 April 2008
© Springer Science + Business Media B.V. 2008

**Abstract** We present the preliminary results of a provenance study of obsidians samples from Cerrillos (ca. 800–100 B.C.) using Mössbauer Spectroscopy. The Cerrillos archaeological site, located in the Upper Ica Valley, Peru, is the only Paracas ceremonial center excavated so far. The archaeological data collected suggest the existence of a complex social and economic organization on the south coast of Peru. Provenance research of obsidian provides valuable information about the selection of lithic resources by our ancestors and eventually about the existence of communication routes and exchange networks. We characterized 18 obsidian artifacts samples by Mössbauer spectroscopy from Cerrillos. The spectra, recorded at room temperature using different velocities, are mainly composed of broad asymmetric doublets due to the superposition of at least two quadrupole doublets corresponding to $Fe^{2+}$ in two different sites (species A and B), one weak $Fe^{3+}$ doublet (specie C) and magnetic components associated to the presence of small particles of magnetite. Multivariate statistical analysis of the Mössbauer data (hyperfine parameters) allows to defined two main groups of obsidians, reflecting different geographical origins.

**Keywords** Obsidians · Mössbauer spectroscopy · Magnetite · Cerrillos—Peru

## 1 Introduction

Obsidian artifacts are a useful tool to determine relationships between prehispanic populations, as several geological sources of obsidian have been located in the Central

A. Bustamante (✉)
Facultad de Ciencias Físicas, Universidad Nacional Mayor de San Marcos, Apartado Postal 14-0149,
Lima 14, Peru
e-mail: abustamanted@unmsm.edu.pe

M. Delgado
Qallta, Jr. Luis Romero 1065-Urb. Roma, Lima 01, Peru

R. M. Latini · A. V. B. Bellido
Instituto de Química, Depto. Físico-Química, UFF, Rio de Janeiro, Brazil

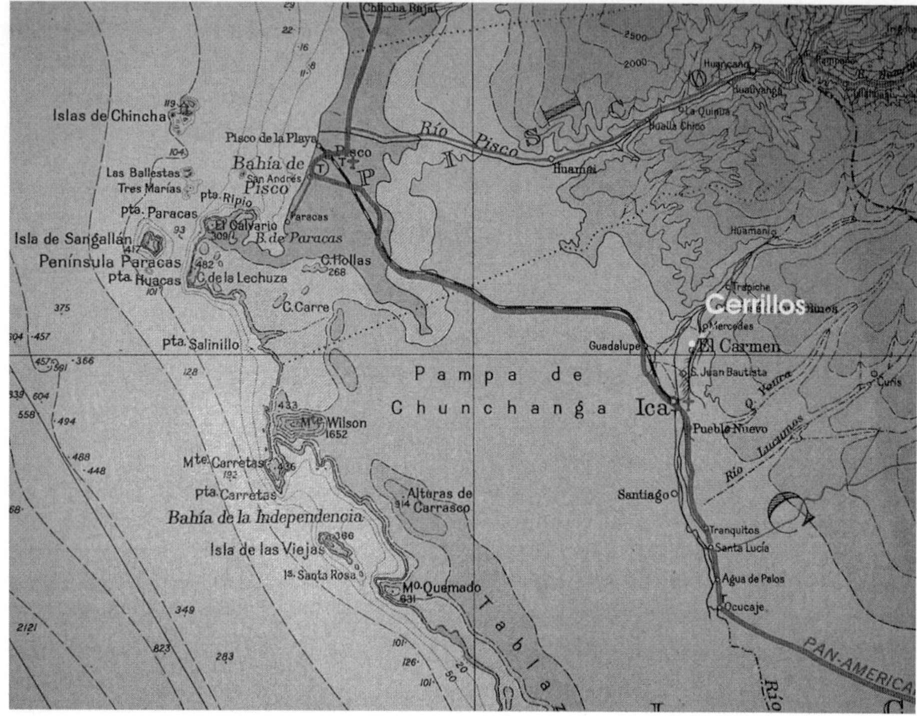


**Fig. 1** Location of geological obsidian sources and location of the Cerrillos archaeological site

Andes (Fig. 1). Data obtained from chemical and physical characterization provide the most reliable evidence of such contacts.

Characterization studies of obsidian artifacts and the determination of their potential geological sources can help archaeologists to better understand Peruvian prehispanic cultures. Although Peru has the most important evidence of cultural development in South America, only few studies focus on the existence of interchange and trade routes.

This study is based on one of several methods used to characterize obsidians from the Central Andes. A. Bustamante and M. Delgado and several researchers from San Marcos University, have initiated a long-term research project to create a database of Peruvian obsidians. The Peruvian museums have very important collections of obsidians artifacts, all of them with known provenance. This is a valuable cultural resource. Its study may help to better understand the interactions between prehispanic human groups in the Andean area.

The studied obsidian samples come from Cerrillos, an early-middle Paracas ceremonial center (ca. 800–100 B.C.), located in the Upper Ica Valley, Peru. This site is a good example of monumental architecture showing evidence of the existence of a complex social organization on the south coast of Peru. Excavations showed that the site has been extensively modified over the centuries. In some places, the archaeological strata are four meters deep, revealing a series of five stratified terrace constructions, and indicating that the site held importance for a much longer period than most of the know Paracas sites. Obsidian artifacts have been mostly recovered from architectural fill deposited throughout many renovations.

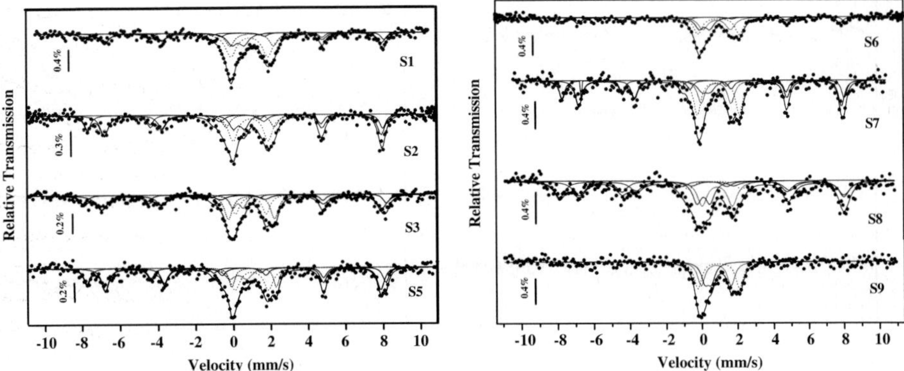

Fig. 2 Mössbauer spectra of samples S1 through S9

Obsidian was the preferred material for making knives, lance and arrow heads and many other tools throughout the prehispanic world due to its fracturing, knapping and flaking qualities. Since volcanic flows are very homogenous, the element profiles are characteristic of a given source. On the other hand, obsidian is stable against chemical alteration, so an analysis of an obsidian artifact reliably reflects the composition of the original flow [1].

## 2 Experimental

Eighteen samples of obsidians were analyzed by Mössbauer Spectroscopy at room temperature (RT) in standard transmission geometry. Measurements were taken with two spectrometers operating in sinusoidal and triangular mode, respectively. Calibration was done using an alpha-Fe foil and isomer shifts are given relative to metallic iron. Source and absorber were kept at the same temperature. The NORMOS code [2] was used for the spectrum analysis. The Statistical Analysis System (SAS) was used to perform multivariate analyses.

## 3 Results and discussion

The Mössbauer spectra (Figs. 2 and 3) exhibit a typical broad asymmetrical doublet, similar to those already observed by others authors [3–7] and due to the superposition of at least two quadrupole doublets associated to $Fe^{2+}$ in two sites with different degrees of distortion (species A and B) and a weaker $Fe^{3+}$ doublet (species C). These doublets are mostly assigned to iron dissolved in the silicate glass phase and having different oxidation states and coordinations. In some samples there is a rather large magnetic contribution due to the presence of magnetite.

The Mössbauer parameters were used as variables in the statistical treatment. The multivariate statistical analysis SPSS packet was used for the classification and ordination. For the cluster analysis methods initially, the data were normalized using "$z$-scores" a transformation that generated new variables with an average of zero and standard deviation equal to one. The Euclidean Distance as a measure of dissimilarity in $n$-dimensional space, and the hierarchical means in the Ward subroutine (minimum variance) was used. The great

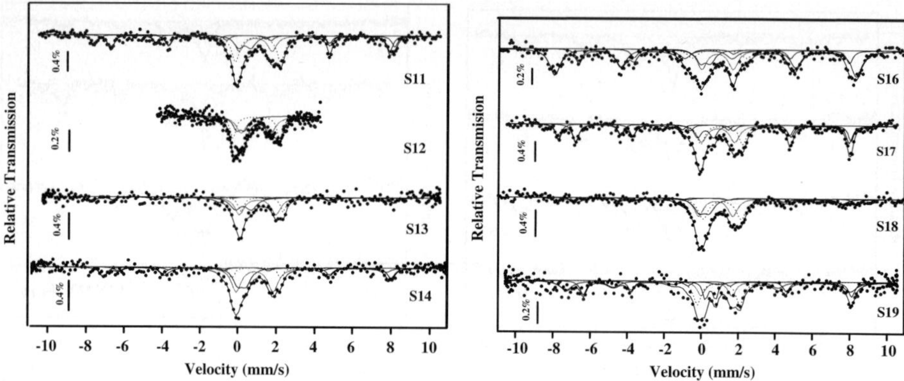

Fig. 3 Mössbauer spectra at RT of obsidian samples S11 through S19. Sample S12 does not contain a magnetic phase

Table 1 Mössbauer parameters of obsidian samples from Cerrillos, Valle de Ica, Peru. Isomer shifts (IS), quadrupole splittings (QS) and line widths (W) are given in mm/s, A(Mag) is given in %

| No. | IS (A) | QS (A) | W (A) | A (A) | IS (B) | QS (B) | W (B) | A (B) | IS (C) | QS (C) | W (C) | A (C) | $Fe^{3+}/Fe^{2+}$ | A (Mag) |
|---|---|---|---|---|---|---|---|---|---|---|---|---|---|---|
| S1 | 1.170 | 2.33 | 0.54 | 16 | 1.032 | 1.96 | 0.7 | 36 | 0.43 | 0.64 | 0.6 | 14 | 0.269 | 34 |
| S2 | 1.059 | 2.57 | 0.47 | 10 | 1.071 | 1.87 | 0.57 | 25 | 0.459 | 0.6 | 0.5 | 11 | 0.314 | 54 |
| S3 | 1.101 | 2.46 | 0.51 | 23 | 1.06 | 1.64 | 0.5 | 18 | 0.376 | 0.55 | 0.42 | 9 | 0.22 | 50 |
| S5 | 1.214 | 2.38 | 0.36 | 12 | 1.085 | 1.79 | 0.62 | 22 | 0.358 | 0.9 | 0.7 | 12 | 0.353 | 54 |
| S6 | 1.139 | 2.32 | 0.55 | 30 | 1.033 | 1.54 | 0.6 | 24 | 0.346 | 0.6 | 0.55 | 19 | 0.352 | 27 |
| S7 | 1.100 | 2.35 | 0.48 | 25 | 0.978 | 1.67 | 0.48 | 16 | 0.427 | 0.45 | 0.65 | 9 | 0.22 | 50 |
| S8 | 0.884 | 2.98 | 0.65 | 7 | 1.008 | 1.69 | 0.65 | 20 | 0.235 | 0.73 | 0.59 | 15 | 0.556 | 58 |
| S9 | 1.045 | 2.33 | 0.5 | 39 | 0.904 | 1.67 | 0.5 | 35 | 0.457 | 0.37 | 0.5 | 26 | 0.351 | 0 |
| S11 | 1.120 | 2.36 | 0.51 | 28 | 0.997 | 1.7 | 0.57 | 18 | 0.435 | 0.64 | 0.6 | 14 | 0.304 | 40 |
| S12 | 1.052 | 2.43 | 0.4 | 34 | 1.037 | 1.54 | 0.65 | 42 | 0.227 | 0.42 | 0.5 | 24 | 0.316 | 0 |
| S13 | 1.266 | 2.21 | 0.35 | 17 | 1.032 | 1.92 | 0.6 | 35 | 0.274 | 0.85 | 0.7 | 27 | 0.519 | 21 |
| S14 | 1.090 | 2.53 | 0.32 | 9 | 1.033 | 1.88 | 0.61 | 36 | 0.331 | 0.47 | 0.59 | 23 | 0.511 | 32 |
| S16 | 1.037 | 2.93 | 0.65 | 8 | 0.939 | 1.69 | 0.63 | 17 | 0.292 | 0.53 | 0.6 | 10 | 0.4 | 65 |
| S17 | 1.189 | 2.26 | 0.5 | 19 | 0.995 | 1.78 | 0.65 | 21 | 0.31 | 0.87 | 0.5 | 12 | 0.3 | 48 |
| S18 | 1.230 | 2.05 | 0.65 | 31 | 0.886 | 1.85 | 0.6 | 25 | 0.262 | 0.53 | 0.68 | 20 | 0.357 | 24 |
| S19 | 1.066 | 2.31 | 0.52 | 23 | 0.948 | 1.84 | 0.5 | 14 | 0.626 | 0.55 | 0.32 | 11 | 0.297 | 52 |
| S20 | 1.025 | 2.65 | 0.6 | 25 | 1.065 | 1.62 | 0.7 | 46 | 0.276 | 0.75 | 0.45 | 9 | 0.127 | 20 |
| S21 | 1.079 | 2.3 | 0.52 | 29 | 1.005 | 1.56 | 0.35 | 8 | 0.533 | 0.62 | 0.32 | 9 | 0.243 | 54 |

advantage of this method is that a priori do not need to make any assumption about the membership of the individual groups.

Table 1 list the data used in the statistical analysis. As a result of the cluster analysis the studied samples were divided into two groups (Fig. 4). The characterization of each one of the defined groups was made by the average and standard deviation of each one of the studied parameters (Table 2)

The discriminates variables by internal correlations for the defined groups explained the cluster analysis output are: Group A variables A(A); A(B); A(C); $Fe^{+3}/Fe^{+2}$; W(B). Group B QS(A); IS(A); W(A) e QS(B).

 Springer

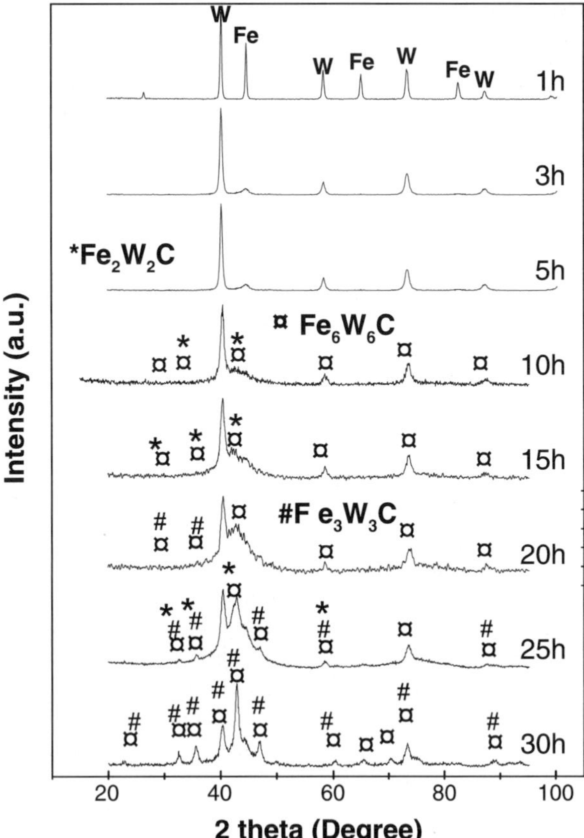



**Fig. 2** X-ray diffraction pattern sequence for Fe, W, and C milled powders at 1, 3, 5, 10, 15, 20, 25 and 30 h of milling

up to 5 milling hours W and Fe peaks diminished, indicating that the alloy of the elemental powders (Fe, W and C) has begun. W and Fe lattices are both bcc, however their atomic radii are too different and, tungsten practically does not dissolve into the iron lattice at room temperature. It is well known, however, that by means of the mechanical alloying technique, it is possible to extend the Solid solubility limits; besides the graphite acts as well as a lubricant and a process control agent. This fact allows forming a true alloy in a short time.

From 10 and up to 25 milling hours, ternary carbides are exhibited: Fe$_3$W$_3$C, Fe$_6$W$_6$C and Fe$_2$W$_2$C, the last one and according to the equilibrium diagram, is unstable and its grain size diminish continuously. At thirty hours the ternary carbide Fe$_2$W$_2$C has disappeared completely and the carbides Fe$_3$W$_3$C and Fe$_6$W$_6$C, which are the stable phases, are much consolidated.

Figure 3 shows the $^{57}$Fe Mössbauer spectra and the corresponding hyperfine field distribution for the milled samples during the times consider above, the hyperfine parameters are listed in Table 1. The spectra were fitted by using MOSFIT program [11] Mössbauer spectra show that during the first milling hour, a sextet characteristic of pure iron is present with a magnetic field of 32.92 T, Alloy does not take place, and this is in agreement with XRD pattern. At three hours of milling the Mössbauer spectrum fit shows five magnetic and two paramagnetic sites. These paramagnetic sites indicate that the ternary

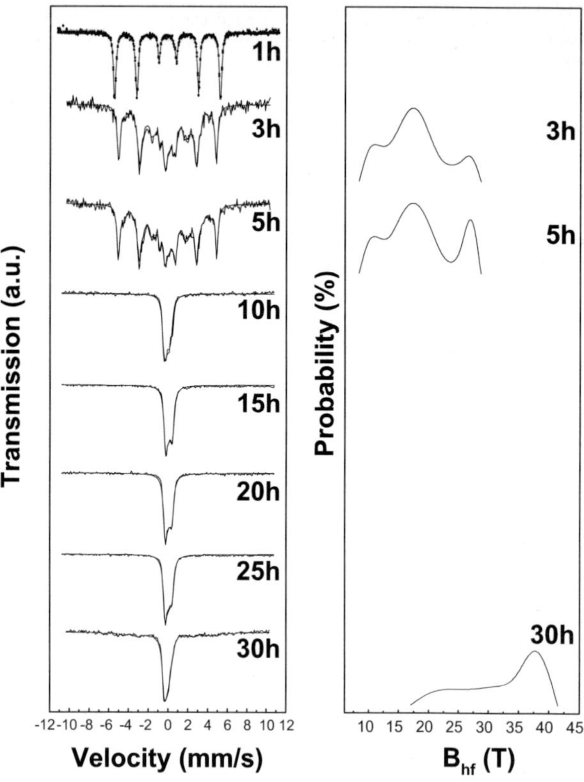

**Fig. 3** Mössbauer spectra and their corresponding hyperfine field distribution for the milled samples during the different times



carbides are beginning to form. The magnetic sites are indicating that tungsten is dissolving into the iron and the paramagnetic sites that the iron is dissolving into tungsten. These facts are not exhibited with XRD. When one W atom is located in the first neighbor shell of a Fe nucleus, the iron hyperfine field decreases by $4.10\pm0.42\,T$ on average and if it is located in the second neighbor shell, the iron hyperfine field of diminish by $2.60\pm0.14\,T$ on average [12]. At five milling hours the paramagnetic phases continued consolidating. At 10 milling hours the spectrum fit shows only two paramagnetic phases, that correspond to $Fe_2W_2C$ and $Fe_6W_6C$ ternary carbides. During the following milling times the phase $Fe_2W_2C$ continues its transformation, giving place to the formation of the phase $Fe_3W_3C$ and even consolidating the phase $Fe_6W_6C$. At the 30 h two paramagnetic phases are present the $Fe_6W_6C$ and $Fe_3W_3C$ ternary carbides. The Mössbauer spectrum fitting also shows a magnetic site with a hyperfine field of 37.7 T. This magnetic site could correspond to a tungsten iron oxide; that could be formed when the jars are opened to take out the powder samples at the different milling times. This phase is present in a small quantity and it is detected by XRD, after a Rietveld refinement, in which it could be read; 43.63 wt % $Fe_3W_3C$, 55.27 wt% $Fe_6W_6C$, ant the balance small quantities of tungsten and possible tungsten and iron oxides. Vickers microhardness measurements of 30 h milled sample was conducted at room temperature with a load of 0.245 N for 20 s; The average was 1564 Vickers (15.6 GPa). This value is higher than that reported elsewhere for the tungsten carbide. In Fig. 4 the prints are shown.

**Table 1** Hyperfine parameters for the milled samples at different times

| Milling time (h) | Sites | $\delta$ (mm/s) | $\Delta E_Q$ (mm/s) | $B_{HF}$ (T) |
|---|---|---|---|---|
| 01 | SI | 0.219 | −0.015 | 32.92 |
| 03 | S1 | 0.188 | −0.063 | 30.40 |
|  | S2 | 0.530 | −0.337 | 23.62 |
|  | D1 | 0.114 | 0.844 |  |
|  | D2 | 0.192 | 0.321 |  |
|  | MD |  |  |  |
|  | S3 | 0.076 | −0.275 | 10.8 |
|  | S4 | 0.064 | −0.275 | 17.2 |
|  | S5 | 0.040 | −0.275 | 26.5 |
| 05 | S1 | 0.182 | −0.055 | 30.65 |
|  | S2 | 0.554 | −0.249 | 25.38 |
|  | D1 | 0.049 | 1.036 |  |
|  | D2 | 0.133 | 0.321 |  |
|  | MD |  |  |  |
|  | S3 | 0.076 | −0.275 | 10.9 |
|  | S4 | 0.064 | −0.275 | 17.3 |
|  | S5 | 0.040 | −0.275 | 26.9 |
| 10 | D1 | 0.002 | 0.594 |  |
|  | D2 | 0.257 | 0.434 |  |
| 15 | D1 | 0.002 | 0.594 |  |
|  | D2 | 0.069 | 0.203 |  |
| 20 | D1 | 0.086 | 0.594 |  |
|  | D2 | 0.030 | 0.326 |  |
| 25 | D1 | 0.086 | 0.594 |  |
|  | D2 | −0.047 | 0.311 |  |
| 30 | D1 | 0.169 | 0.594 |  |
|  | D2 | −0.079 | 0.342 |  |
|  | MD |  |  |  |
|  | S1 | −0322 | 0.250 | 37.7 |

$\delta$ data are given relative to $\alpha$-Fe.

*S* sextet, *D1* doublet 1, *D2* doublet 2, *MD* magnetic distribution

**Fig. 4** Optical micrograph showing the microhardness prints on a particle of ternary carbide for a sample of 30 h of mill

## 4 Conclusions

The results of this study show that Mechanical alloying is a technique by which it is possible to extend the solid solubility limits. With the milling conditions, and at 30 h, the $Fe_6W_6C$ and $Fe_3W_3C$ carbide were obtained in spite of the almost null solubility of W and Fe, at room temperature. These ternary carbides are paramagnetic phases as revealed by Mössbauer fit. They are very hard phases with microhardness of 15.6 GPa.

**Acknowledgement** This Work has been carried out with Colciencias, The Universidad del Valle and The Universidad Santiago de Cali. The authors are greatly indebted for financial support.

## References

1. Suryanarayana, C.: Prog. Mater. Sci. **46**, 1–184 (2001)
2. Jartych, E., et al.: J. Phys., Condens. Matter **10**, 4929 (1998)
3. Jartych, E., et al.: J. Magn. Mater. **208**, 221 (2000)
4. Minamino, Y., Koizumi, Y., Tsuji, N., Hirohata, N., Mizuuchi, K., Ohkanda, Y.: Sci. Technol. Adv. Mater. **5**, (2), 133–143 (2004)
5. ASM Metals Handbook Vol. 03—Alloy phase diagramans
6. Goldschmidt, H.J.: Interstitial Alloys. Plenum, New York (1967)
7. Zak, T., et al.: J. Magn. Magn. Mater. **272–276**, 1119–1121 (2004)
8. Novakova, A.A., et al.: J. Alloys Comp. **317–318**, 423–427 (2001)
9. Zhang, Z.W., et al.: J. Alloys Comp. **370**, 186–191 (2004)
10. Ferrari, M., Lutteroti, L.J.: J. Appl. Phys. **72**, 7246 (1194). http://www.ing.unitn.it/~kytteri/maud
11. Varret, F., Teillet, J.: Unpublished MOSFIT program, Université du Maine, France.
12. Jartych, E., et al.: J. Magn. Magn. Mater. **272–276**, 1119–1121 (2004)

Hyperfine Interact (2007) 175:55–61
DOI 10.1007/s10751-008-9588-x

# Iron-containing pyrochlores: structural and magnetic characterization

C. K. Matsuda · R. Barco · P. Sharma · V. Biondo ·
A. Paesano Jr. · J. B. M. da Cunha · B. Hallouche

Published online: 26 March 2008
© Springer Science + Business Media B.V. 2008

**Abstract** In the present study, iron-containing pyrochlores of the $A_2FeBO_7$ type (where A=Y, Gd, Eu and Dy; B=Sb and Nb) were synthesized by solid-state reaction of precursor oxides. The compounds were structurally and magnetically characterized by X-ray diffraction, Mössbauer spectroscopy and magnetic measurements. The results revealed that the crystallographic structures and magnetic properties depend, basically, on the type of B atoms.

**Keywords** Pyrochlores · Iron · Mössbauer spectroscopy

## 1 Introduction

Pyrochlores are oxides with the ideal formula $A_2B_2O_7$, where $A$ and $B$ are in general trivalent and tetravalent cations, respectively. These compounds are predominantly cubic and ionic in nature, and represent a family of phases isostructural to the mineral pyrochlore $(NaCa)\cdot(NbTa)O_6F/(OH)$. The $A_2B_2O_7$ compounds exhibit a wide variety of interesting physical properties, the most striking one being the magnetic frustration [1, 2].

It is possible to synthesize these compounds through chemical substitutions at the A and B sites, provided the ionic radius and charge neutrality criteria are satisfied [3]. In this sense, only a few iron containing pyrochlores of the type $A_2(FeB)O_7$, where B is a pentavalent cation, were previously synthesized [3]. In addition, Mössbauer spectroscopy has been scarcely applied to study this important class of oxides, principally regarding to

C. K. Matsuda · R. Barco · P. Sharma · V. Biondo · A. Paesano Jr. (✉)
Departamento de Física, UEM, Av. Colombo, 5790, 87.020-900 Maringá, Paraná, Brazil
e-mail: paesano@wnet.com.br

J. B. M. da Cunha
Instituto de Física, UFRGS, Porto Alegre, Rio Grande do Sul, Brazil

B. Hallouche
Departamento de Química e Física, UNISC, Santa Cruz do Sul, Rio Grande do Sul, Brazil

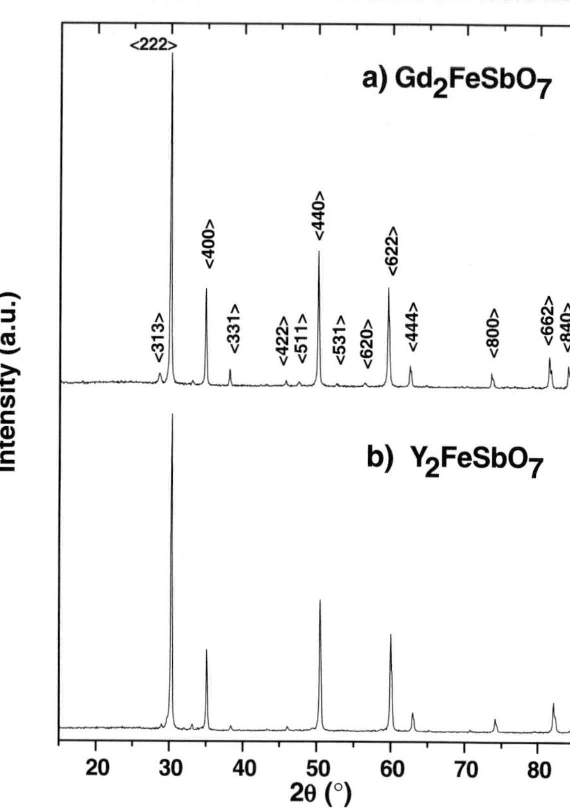



**Fig. 1** X-ray patterns for the Gd/Sb (**a**) and Y/Sb (**b**) samples

low temperature studies. Some choices for the B atom were not examined yet, even at room temperature (RT).

In this sense, in this work we prepared and characterized iron-containing pyrochlores, extending the number of the iron partners in the general formula $A_2$(Fe$B$)O$_7$.

## 2 Experimental

Iron-containing pyrochlores of the $A_2$FeBO$_7$ type (where A=Y, Gd, Eu and Dy; B=Sb and Nb) were prepared by annealing mixtures of $A_2O_3$, $Fe_2O_3$ and $Nb_2O_5$ or $Sb_2O_3$ powders (purity of 99.9%), in appropriate molar ratios. Before the heat treatment, performed under oxygen atmosphere at 1100°C for 24 h, the precursors were milled for 3 h in a vial of alumina with zirconia rods, under argon atmosphere, using a planetary ball mill (Fritsch – pulverisette 6). The ball-to-powder mass ratio was 20:1 and the speed of rotation was 300 rpm.

The compounds, henceforth designated by A/B, were structurally and magnetically characterized by X-ray diffraction, Mössbauer spectroscopy and magnetic measurements. X-ray diffraction measurements were performed using Cu radiation, in a Shimadzu-6000 Diffractometer, in the conventional θ–2θ Bragg–Brentano geometry. Mössbauer spectra were taken from a constant acceleration spectrometer with a $^{57}$Co(Rh) source, using absorbers with about 10 mg(Fe)/cm$^2$. For the low temperature Mössbauer measurement, a

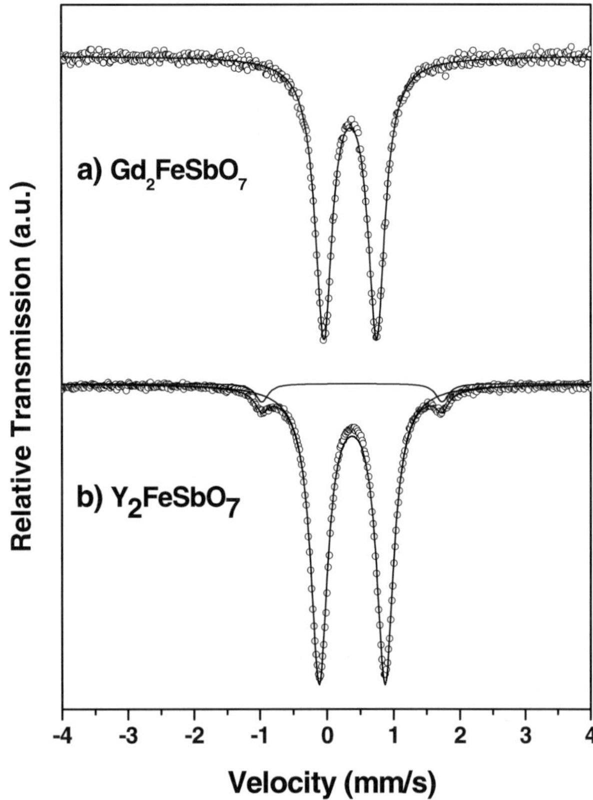

**Fig. 2** RT Mössbauer spectra for the Gd/Sb (**a**) and Y/Sb (**b**) samples



Janis (SVT-400) liquid helium cryostat was used. The magnetic properties of the randomly pressed powders were measured in a vibrating sample magnetometer, with a maximum field of 1.5 T.

## 3 Results and discussions

Two representative diffractograms for the A/Sb series are shown in the Fig. 1. Apart from small differences in the peak positions and intensities, reflecting the expected variations in the lattice parameters and scattering factors, both compounds (in fact, all for this series) reveal the same crystallographic structure of the ternary pyrochlores (i.e., the Fd3m symmetry).

Figure 2 shows the RT Mössbauer spectra for the same two samples. The main contribution, and unique for the Gd/Sb compound, is a doublet. The fitted values for the quadrupole splittings of the antimony series (Doublets S) are in very good agreement with earlier reported values [4] (a compilation of the Mössbauer parameters from all the measured spectra is given in Table 1). On the other hand, in the Y/Sb spectrum there is visible an additional quadrupole doublet (Doublet M) of small intensity with a larger splitting. The explanation for this uncommonly large splitting is still lacking.

The Mössbauer spectrum for the Gd/Sb obtained at 5 K is shown in Fig. 3. It reveals a well resolved magnetic hyperfine splitting. According to Knop et al. [4], this was not

**Table 1** Hyperfine parameters and subspectral areas for the pyrochlore samples

| Pyrochlore | Subspectrum | $IS^a$ (mm/s) | QS (mm/s) | $B_{hf}$ (kOe) | $\Gamma$ (mm/s) | Area (%) |
|---|---|---|---|---|---|---|
| RT | Doublet S | 0.36 | 0.79 | – | 0.31 | 100.0 |
| $Gd_2FeSbO_7$ | | | | | | |
| 5K | Sextet | 0.54 | −0.07 | 483 | 1.10 | 100.0 |
| $Eu_2FeSbO_7$ | Doublet S | 0.38 | 0.75 | – | 0.32 | 100.0 |
| $Dy_2FeSbO_7$ | Doublet S | 0.38 | 0.93 | – | 0.30 | 100.0 |
| $Y_2FeSbO_7$ | Doublet S | 0.38 | 1.10 | – | 0.31 | 95.5 |
| | Doublet M | 0.39 | 2.70 | – | 0.23 | 4.5 |
| $Gd_2FeNbO_7$ | Sextet | 0.30 | −0.15 | 493 | 0.56 | 27.1 |
| | Doublet I | 0.39 | 0.97 | – | 0.46 | 42.2 |
| | Doublet E | 0.24 | 2.30 | – | 0.40 | 30.7 |
| $Eu_2FeNbO_7$ | Sextet | 0.35 | −0.19 | 504 | 0.57 | 41.6 |
| | Doublet I | 0.38 | 0.94 | – | 0.43 | 32.7 |
| | Doublet E | 0.26 | 2.22 | – | 0.39 | 25.7 |
| $Dy_2FeNbO_7$ | Sextet | 0.32 | 0.04 | 556 | 1.00 | 22.1 |
| | Doublet I | 0.42 | 1.01 | – | 0.43 | 47.1 |
| | Doublet E | 0.25 | 2.84 | – | 0.38 | 30.8 |
| $Y_2FeNbO_7$ | Sextet | 0.27 | −0.30 | 480 | 1.09 | 14.7 |
| | Doublet I | 0.37 | 1.03 | – | 0.43 | 47.8 |
| | Doublet E | 0.25 | 2.69 | – | 0.43 | 37.5 |

*IS* Isomer shift, *QS* Quadrupole splitting, $B_{hf}$ hyperfine magnetic field;

$^a$ Relative to $\alpha$–Fe

observed for this pyrochlore down to 25.4 K. The spectrum of Fig. 3 was fitted with a sextet although the large linewidth indicates the occurrence of a hyperfine magnetic field distribution. The obtained hyperfine magnetic field of 483 kOe is significantly larger than that previously reported for the Y/Sb case with 400 kOe [5].

X-rays diffractograms for two selected A/Nb samples are presented in Fig. 4. The results for the Eu/Nb and Y/Nb (not shown) closely reproduce the same pattern. Evidently, the patterns are different to those of the A/Sb series. A careful examination of the X-ray profiles gives no clear evidence for un-reacted precursors or perovskite-like phases. These results indicate that the crystalline structure for the A/Nb series is different from that represented by the Fd3m symmetry. Rhombohedral distortions have been reported for pyrochlore systems such as $RE_2Fe_{1.33}W_{0.67}O_7$ (p. 99 in [3]), although no diffractograms are presented. Aiming to find out the actual structure, we are currently carrying out a Rietveld analysis of the data obtained from the A/Nb systems.

Figure 5 shows the RT Mössbauer spectra for the $Gd_2FeNbO_7$ and $Dy_2FeNbO_7$ samples. All spectra of this series of samples were fitted with three components: two doublets (the inner=Doublet I; the outer=Doublet E) and one discrete sextet, the large line width of which is also indication of a magnetic distribution. According to this, up to three iron sites could be assigned for the A/Nb pyrochlore structure. Interestingly, one of them is magnetic at RT whereas the other two are not. Within a few percent of error, the areas relation for the non-magnetic sites is 3:2, although the relative area for the magnetic site shows a strong variation from one sample to other.

Figure 6 presents the magnetization at RT, as a function of the applied field, for the A/Sb compounds. As expected for paramagnetic samples, a linear behavior was obtained for all samples. However, hysteretic contributions are also revealed in the magnetization curves, which are easily visible for the Gd/Sb and more faintly for the other pyrochlores. We have

**Fig. 3** Mössbauer spectra taken at 5 K for the Gd/Sb sample

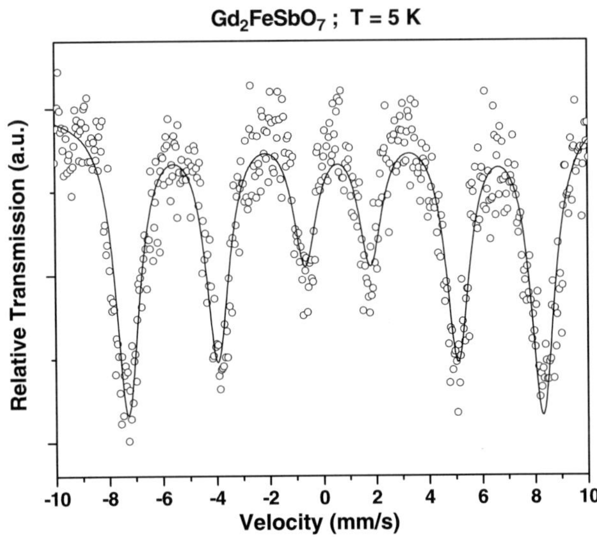


**Fig. 4** X-ray patterns for the Gd/Nb (**a**) and Dy/Nb (**b**) samples

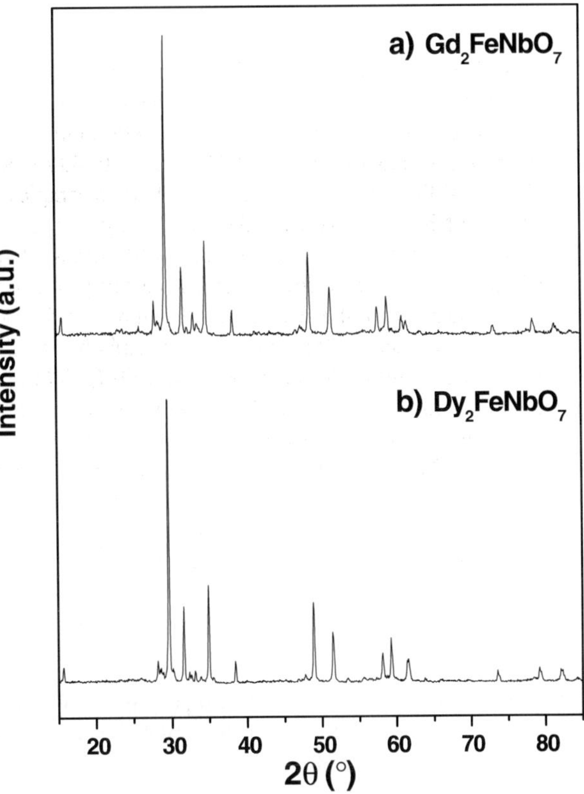

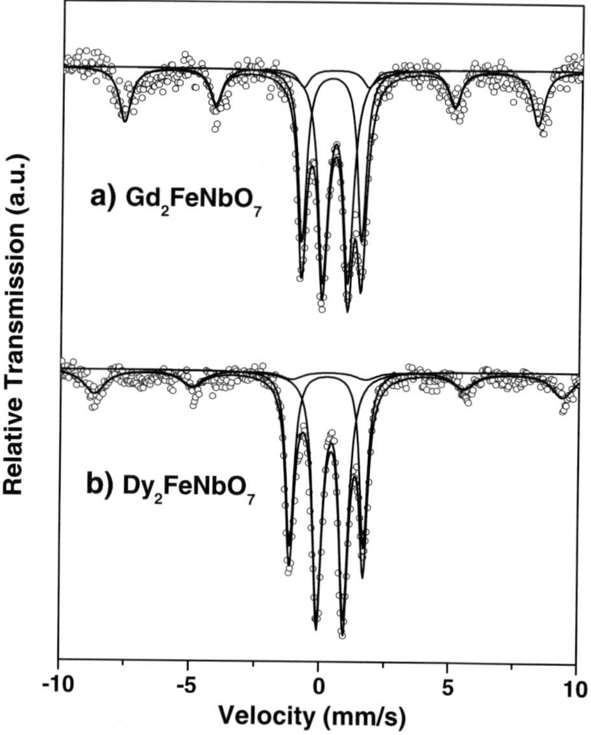

Fig. 5 RT Mössbauer spectra for the Gd/Nb (a) and Dy/Nb (b) samples



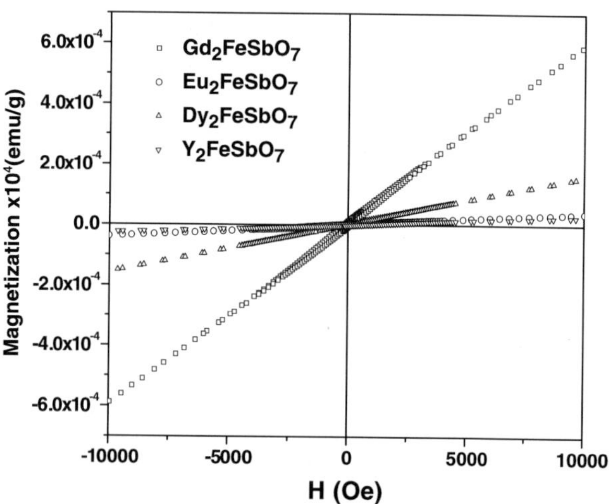

Fig. 6 RT magnetization curves, as a function of the applied field, for the A/Sb pyrochlores



no definitive explanation for this fact, since there is no indication from the Mössbauer data for a phase containing magnetic iron in any of the A/Sb pyrochlores. In contrast, the hysteretic contributions in the magnetization curves of the A/Nb systems, as shown in Fig. 7, can be related to the magnetic contribution also seen in the Mössbauer spectra.

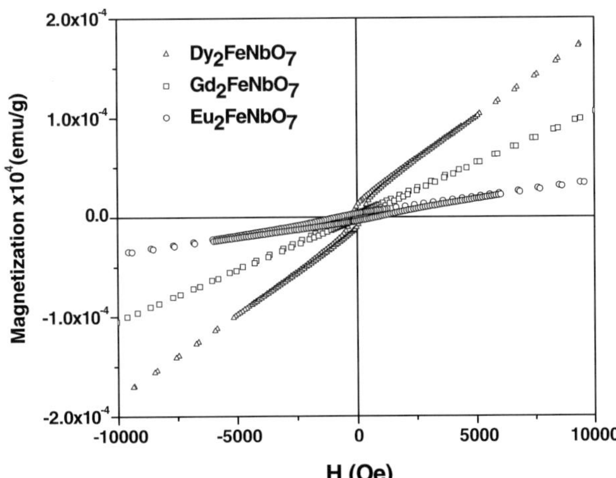



**Fig. 7** RT magnetization curves, as a function of the applied field for the A/Nb pyrochlores

When comparing the magnetic behavior of both systems (i.e., A/Sb and A/Nb), no evident influence from the A element can be identified.

## 4 Conclusions

The resulting crystallographic structure for $A_2FeBO_7$ pyrochlores depends on the B chemical species present in the compound. When B is antimony, the structure retains its original cubic symmetry, with iron occupying one or, at maximum, two crystalline non-cubic sites. All the compounds are paramagnetic at RT but the Gd/Sb shows magnetic order at 5 K. If B is niobium, the structure changes the crystallographic symmetry. Two or three sites are possible for iron occupation, one of them being magnetic at RT.

**Acknowledgements** The authors would like to thank to the Brazilian agencies, CNPq (Bolsa PDJ) and CAPES, for financial support.

## References


1. Greedan, J.E.: J. Alloys Compd. **408412**, 444 (2006)
2. Greedan, J.E.: J. Mater. Chem. **11**, 37 (2001)
3. Subramanian, M.A., Aravamudan, G., Subba Rao, G.V.: Prog. Solid St. Chem. **15**, 55 (1983)
4. Knop, O., Brisse, F., Meads, R.E., Bainbridge, J.: Can. J. Chem. **46**, 3829 (1968)
5. Knop, O., Brisse, F., Castelliz, L., Sutarno, R.: Can. J. Chem. **43**, 2812 (1965)

Hyperfine Interact (2007) 175:63–70
DOI 10.1007/s10751-008-9589-9

# Study of the effect of Mn and Cu in Fe–Mn–Al–C–Cu alloys by ICEMS and XRD

**J. D. Betancur-Ríos · J. A. Tabares ·
G. A. Pérez Alcázar · V. F. Rodríguez**

Published online: 4 April 2008
© Springer Science + Business Media B.V. 2008

**Abstract** Experimental analysis of magnetic and structural properties of Fe–Mn–Al–C–Cu alloys with compositions $Fe_xMn_{0.915-x}Al_{0.075}C_{0.01}$ (series A) and $Fe_xMn_{0.912-x}Al_{0.075}C_{0.01}$ $Cu_{0.003}$ (series B), $0.500 \leq x \leq 0.800$, in steps of 0.050 is presented and discussed. The analysis was performed by integral conversion electrons Mössbauer spectrometry and X-ray diffraction at room temperature. The results suggest, for both series of alloys, that for the highest Mn content, samples exhibit an antiferromagnetic behavior, typical of the FCC or austenite FeMn phase rich in Mn; for those of low Mn content, the coexistence of paramagnetic austenite, typical of the FeMn alloy poor in Mn, a ferromagnetic BCC or ferrite phases can be observed, while for the lowest Mn content, only ferromagnetic (FM) phase tends to prevail. The FM phase is associated to the BCC FeMnAl as was corroborated by X-ray diffraction. The samples with the highest Mn content, the influence of Cu addition is to reduce the mean hyperfine field and to stabilize the antiferromagnetic behavior.

**Keywords** Fe–Mn–Al–C alloys · Mössbauer spectrometry · XRD

## 1 Introduction

Alloys based on the Fe–Mn–Al system (ferritic, austenitic or both) are good candidates to replace the conventional stainless steel, which are based on the ternary alloy Fe–Ni–Cr. The Fe–Mn–Al system offers the possibility to replace strategic elements as Ni and Cr by others lighter and cheaper as Mn and Al. Besides, it has attractive mechanical and corrosion resistance properties. Al stabilizes the ferritic phase (BCC) and gives the stainless character, since it mainly produces a passivating layer of $Al_2O_3$ on the surface of the metal; and Mn stabilizes the austenitic phase (FCC), essential to get good mechanical properties at several temperatures [1–3]. Also, C increases the stability of the austenitic phase if its content does not exceed 1%, since more of this amount decreases the oxidation resistance [4]. It had

J. D. Betancur-Ríos (✉) · J. A. Tabares · G. A. Pérez Alcázar · V. F. Rodríguez
Departamento de Física, Universidad del Valle, A. A. 25360 Cali, Colombia
e-mail: dbetan@calima.univalle.edu.co

**Table 1** Chemical composition (in at.%), average austenitic grain size, cell parameter and volume fraction of the obtained phases of all the studied alloys

| Alloy | Fe (at.%) | Mn (at.%) | Al (at.%) | C (at.%) | Cu (at.%) | Austenitic grain size (μm) | Cell parameter (Å) | Phase volume (%) |
|---|---|---|---|---|---|---|---|---|
| A1 | 50 | 41.5 | 7.5 | 1.0 | | 41 | 3.652 | 100 (FCC) |
| A2 | 55 | 36.5 | 7.5 | 1.0 | | 28 | 3.644 | 100 (FCC) |
| A3 | 60 | 31.5 | 7.5 | 1.0 | | 40 | 3.637 | 100 (FCC) |
| A4 | 65 | 26.5 | 7.5 | 1.0 | | 33 | 3.631 | 100 (FCC) |
| A5 | 70 | 21.5 | 7.5 | 1.0 | | 29 | 3.625 | 100 (FCC) |
| A6 | 75 | 16.5 | 7.5 | 1.0 | | 40 | 3.618 (FCC) | 74 (FCC) |
| | | | | | | | 2.898 (BCC) | 26 (BCC) |
| A7 | 80 | 11.5 | 7.5 | 1.0 | | | 3.615 (FCC) | 4 (FCC) |
| | | | | | | | 2.886 (BCC) | 96 (BCC) |
| B2 | 55 | 36.2 | 7.5 | 1.0 | 0.30 | 28 | 3.643 | 100 (FCC) |
| B4 | 65 | 26.2 | 7.5 | 1.0 | 0.30 | 27 | 3.636 | 100 (FCC) |
| B5 | 70 | 21.2 | 7.5 | 1.0 | 0.30 | 26 | 3.632 | 100 (FCC) |
| B6 | 75 | 16.2 | 7.5 | 1.0 | 0.30 | 18 | 3.622 (FCC) | 64 (FCC) |
| | | | | | | | 2.903 (BCC) | 36 (BCC) |
| B7 | 80 | 11.2 | 7.5 | 1.0 | 0.30 | | 3.619 (FCC) | 23 (FCC) |
| | | | | | | | 2.888 (BCC) | 77 (BCC) |

been reported that the optimal weigh composition for which the Fe–Mn–Al–C system presents good properties is 25–30% Mn, 8–10% Al and near 1% C [2–4].

Studies on magnetic properties of bulk Fe–Mn–Al–C system have shown how the magnetic transitions, under cooling, could be classified in accordance with the microstructure: for completely austenitic alloys the transition is from paramagnetic to antiferromagnetic behavior and for duplex alloys (BCC plus FCC) from superparamagnetic to antiferromagnetic behavior [5–7]. The theoretical interpretation of the experimental outcomes of this system, the Mn–Mn and Fe–Mn interactions are considered as antiferromagnetics, the Fe–Fe interaction as weakly ferromagnetic and the Al acts as magnetic dilutor [5–7]. The literature reports several works of the Fe–Mn–Al–C system, but the effect of Mn and Cu addition on its magnetic properties had not been studied by Integral Conversion Electron Mössbauer Spectrometry technique (ICEMS). In these samples, the Cu addition was done in spite of the well known fact that in the weathering steel corrosion resistance is due basically to their lower Cu content (0.3.%) [5].

In this work, a study of the system Fe–Mn–Al–C–Cu was performed by ICEMS and X-ray Diffraction (XRD) at root temperature. The aim was to study the effect of the change of the Mn content and the effect of a small additions of Cu to the specimens. To our best knowledge, there is a lack of work related to the effect of Cu content in this kind of alloys.

## 2 Experimental

In order to obtain the alloys, high purity powders of Fe, Mn, Al, C, and Cu (99.9%) were melted in an arc furnace, in argon atmosphere. The resulted pellets were melted again in an induction furnace and ingots were obtained. The ingots were forged at 1,473 K, laminated at 1,073 K until 2.3 mm, homogenized at 1,323 K during 15 minutes, and finally a fast tempering was done in oil at room temperature. Then, discs of 25 mm diameter and

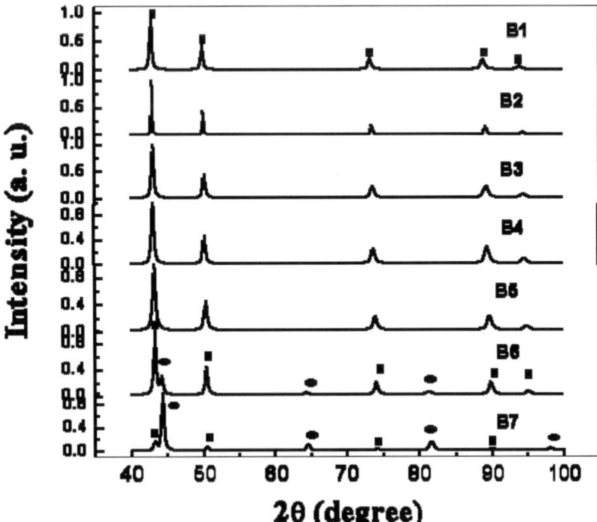



**Fig. 1** X-ray diffraction patterns of samples B. Samples A exhibit the same behavior. *Square* FCC phase, *circle* BCC phase

between 2.0 and 2.5 mm thick were cut [2]. Finally, the discs were grinded by means of different abrasive papers (grain size from 100 to 1,200), polished to mirror brightness with an alumina solution, cleaned ultrasonically in an acetone bath and dried in air.

The stoichiometries of samples were $Fe_xMn_{0.915-x}Al_{0.075}C_{0.01}$ and $Fe_xMn_{0.912-x}Al_{0.075}C_{0.01}Cu_{0.03}$, with $0.500 \leq x \leq 0.800$, in steps of 0.050. The compositions are shown in Table 1 with the following convention: A1 to A7 for samples without Cu, and B2 to B7 for those with Cu. Samples were characterized by ICEMS at room temperature (RT) using a conventional spectrometer in a constant acceleration mode, equipped with a parallel plates avalanche counter (PPAC), a source of $^{57}Co/Rh$ of 20 mCi. Spectra was fitted with Normos–Distri program [8], and all isomer shifts are referred to $\alpha$-Fe. The XRD patterns of the samples were performed at RT using the $Cu/K\alpha$ radiation. To obtain the cell parameter and phase proportion in the case of duplex alloys, it was used MAUD program, which consider Rietveld analysis to refine the XRD patterns [9]. Finally, to obtain the grain size optical microscopy (OM) was used and the ASTM for grain size number [10].

## 3 Results and discussion

### 3.1 XRD

Figure 1 shows the XRD spectra for B alloys (alloys A exhibit the same behavior). The structures of A1 to A5 and B1 to B5 alloys correspond to a FCC structure (austenite phase). These results were expected, mainly due to the high Mn content which favors the austenite formation, and the low carbon content avoids the presence of carbides. The patterns of A6, A7, B6 and B7 alloys exhibited two structures: FCC and BCC, corresponding to austenite and ferrite phases, and in according to the duplex character reported for ternary FeMnAl alloys [11].

After the Rietveld refinement using the MAUD program, the results displayed in Table 1 show that the austenitic proportion in duplex (dual) alloys is lower in those without Cu and that the ferrite volume fraction increases from 26% for A6 alloy to 96% for A7 alloy and

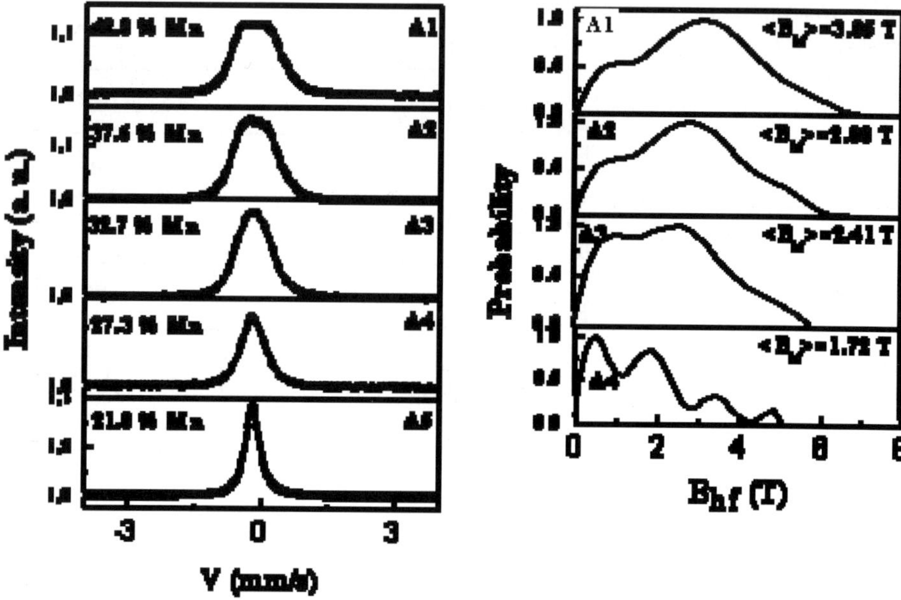

Fig. 2 Mössbauer spectra at RT and their HFDs of samples A1 to A4. Percentages are showed in mass. Sample A5 was fitted with a singlet

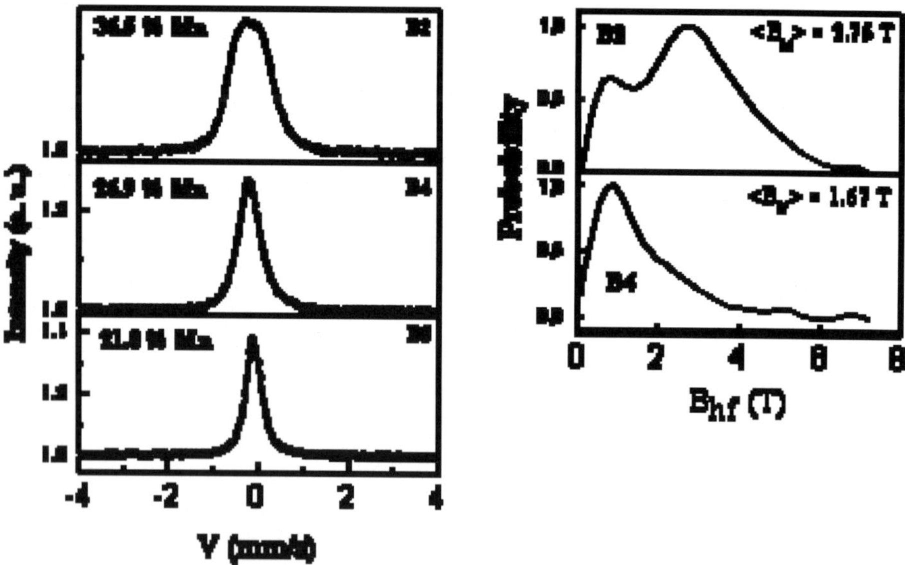

Fig. 3 Mössbauer spectra at RT and their HFDs of B2 and B4 samples. Percentages are showed in mass. Sample B5 was fitted with a singlet

from 36% for B6 alloy to 77% for B7 alloy, suggesting the effect of cupper addition, which is to retain the FCC phase. The results also suggest that when Mn content increases, the austenitic cell grows, which can be explained by the relatively high Mn atomic size related to that of Fe atoms. Additionally, it is observed a small reduction of austenitic grain size

 Springer

**Table 2** Hyperfine parameters and fit model used to fit the obtained ICEMS spectra

| Sample | Component | <IS> (mm/s) | $\Gamma$ (mm/s) | ($\Delta$EQ) (mm/s) | <Bhf> (T) | Area (%) |
|---|---|---|---|---|---|---|
| A1 | HFD | −0.14 | 0.44 | 0.065 | 3.06 | 100 |
| A2 | HFD | −0.15 | 0.44 | 0.060 | 2.81 | 100 |
| A3 | HFD | −0.15 | 0.44 | 0.071 | 2.39 | 100 |
| A4 | HFD | −0.19 | 0.44 | −0.017 | 1.72 | 100 |
| A5 | Singlet | −0.06 | 0.36 | – | – | 100 |
| A6 (+$\alpha$) | Singlet | −0.16 | 0.36 | | | 75 |
| | HFD | −0.19 | – | 0.027 | 27.5 | 25 |
| A7 (+$\alpha$) | Singlet | −0.09 | 0.40 | | | 9 |
| | HFD | −0.18 | – | 0.004 | 28.2 | 91 |
| B2 | HFD | −0.16 | 0.44 | 0.058 | 2.75 | 100 |
| B4 | HFD | −0.26 | 0.44 | −0.116 | 1.67 | 100 |
| B5 | Singlet | −0.06 | 0.36 | – | – | 100 |
| B6 ($\gamma$+$\alpha$) | Singlet | −0.14 | 0.36 | | | 64 |
| | HFD | −0.19 | – | −0.218 | 27.1 | 36 |
| B7 ($\gamma$+$\alpha$) | Singlet | −0.005 | 1.00 | | | 5 |
| | HFD | −0.16 | – | −0.006 | 27.8 | 95 |

(measured by OM) in series B alloys (around 25 µm) in comparison to series A alloys (around 35 µm), which is due to the presence of Cu in these series.

## 3.2 Mössbauer

In Figs. 2 and 3 are shown, the Mössbauer spectra of alloys with high Mn content, A1–A4 and B2–B5, respectively, and their corresponding hyperfine field distributions (HFD). For the fit of the spectra, it was taken into account previous obtained transmission Mössbauer results of binary FeMn FCC disordered alloys reported by Ishikawa and coworkers [12–14] and of ternary FeMnAl FCC disordered alloys with 5.% Al [5]. These binary and ternary systems exhibited broad doublets which seemed as paramagnetic but considering their antiferromagnetic (AF) character, obtained by different techniques, and the disordered character, they where fitted with a HFD and low fields. The AF character is due mainly to Fe–Mn and Mn–Mn interactions. Then for the fit of A1–A4 and B2–B4 samples was considered one HFD, using the Normos–Distri program to fit the spectra [8]. In this case, was observed that the line width and the mean hyperfine field distribution increased with the increase of Mn content; which confirms the AF behavior induced by Mn in these alloys. In the HFD, the most probable sites shift toward lower fields as the Mn content decreases losing the AF character. Comparing these results with those of Cruz et al. [5], can be concluded that the proportion of superficial phases corresponds to the proportion of the present phases in the bulk sample, for the current system. It does suggest that as Mn content decreases the transition could be from AF to paramagnetic (P) behavior, as was reported for Fe–Al–Mn–C austenitic system [5, 15, 16].

Spectra of samples A5 (21.5% Mn) and B5 (21.2.% Mn) exhibited only a P site. They were fitted considering a singlet whose hyperfine parameters are typical of the austenite phase, as is shown in Table 2. In the case of samples A6, A7, B6, and B7, the Mössbauer spectra consisted of a set of ferromagnetic and paramagnetic subspectra related to the BCC and FCC crystal structures, respectively (see Fig. 4 for A samples). Both structures contain substitutional and interstitial atoms randomly distributed [17]. The fitting was carried out by putting one hyperfine field distribution (HFD) for the BCC ferromagnetic part, with a

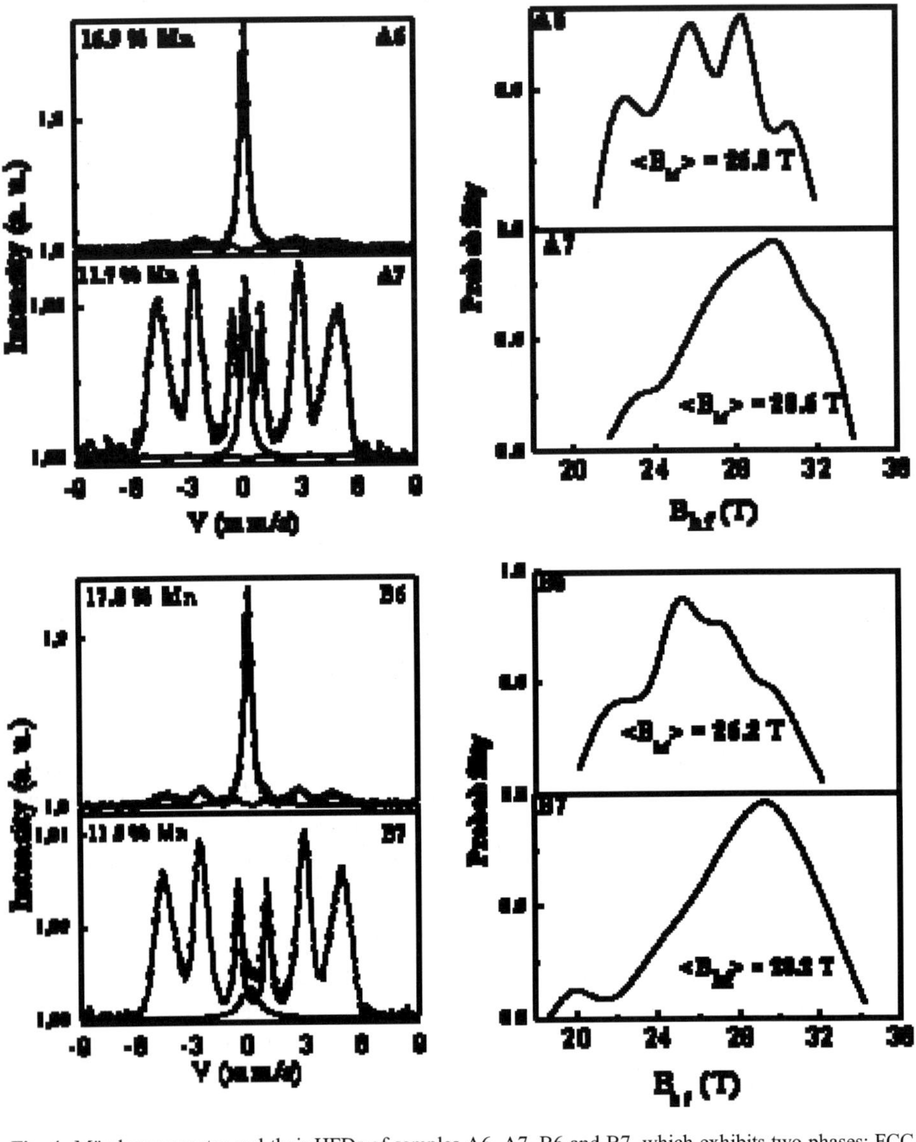

Fig. 4 Mössbauer spectra and their HFDs of samples A6, A7, B6 and B7, which exhibits two phases: FCC and BCC. Percentages are showed in mass

maximum probability characteristic of the FeMnAl ferrite phase [17] and with a paramagnetic site. This paramagnetic site is associated with the FCC structure which now has low Mn content and can not induce the AF character of spectra showed in previous figure for richer Mn alloys. Although in B7 alloy, the P part (5%) is lower, we expected that A7 and B7 alloys were totally FM (disorder BCC phase), due to high iron content. Besides, it can be noted that the width of small singlet in B7 alloy is large, likely due to the presence of impurities and a disorder of the alloy.

If we compare the mean hyperfine fields as a function of Fe concentration among A6, A7, B6 and B7 alloys, we observe a slightly higher mean hyperfine field for the A alloys as a result of the presence of Cu atoms in the B alloys, which are non magnetic and dilute

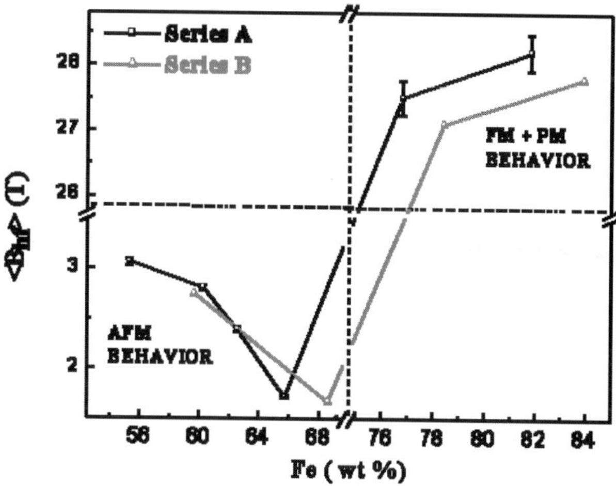

**Fig. 5** Proposed magnetic phase diagram for the studied alloys ($<B_{hf}>$ versus at.% of Fe)

**Fig. 6** Optical micrographies of two of the austenitic samples (*A1* and *A2*)

some ferromagnetic bounds. Another point of view is that the effect of Cu addition is to stabilize the antiferromagnetic phase as it is suggested by the mean hyperfine field of the HFD of alloys B ($<B_{hf}>_{B6}=27.1$ T, $<B_{hf}>_{B7}=27.8$ T) in comparison to alloys A ($<B_{hf}>_{A6}=27.5$ T, $<B_{hf}>_{A7}=28.2$ T). Figure 5 shows these results for the studied alloys in a magnetic phase diagram $<B_{hf}>$ vs. Fe concentration. It can be observed the general reduction of $<B_{hf}>$ of the FCC phase in series B alloys in comparison to those of series A, and the corresponding increases of $<B_{hf}>$ in the BCC phase of the two series.

Finally, in order to illustrate the austenitic phase of the samples, Fig. 6 displays micrographics of samples A1 and A2 that show the typical FCC microstructure with twins and illustrates also the sizes of the grains.

High temperature corrosion experiments in air are now in progress in order to study the oxidation resistance properties of this Fe–Mn–Al–C–Cu system.

# 4 Conclusions

These results suggest that for both alloy series, the one with the highest Mn content, the FCC structure of the samples exhibit an antiferromagnetic behavior, typical of the FCC or

austenite phase. For the series of low Mn content, the coexistence of paramagnetic (austenite) and ferromagnetic (ferrite) phases can be observed and only the FM phase tends to prevail. The ferromagnetic phase is related to ferrite, as was corroborated by X-ray diffraction and Mössbauer spectrometry.

The influence of Cu addition in Fe–Mn–Al–C system causes to reduce the mean hyperfine field and to stabilize the antiferromagnetic phase.

**Acknowledgements** The authors would like to thanks Universidad del Valle and Colciencias for the financial support of this work, and the National Taiwan University of Science and Technology for providing the optical microscopy measurements.

# References


1. Pérez Alcázar, G.A.: Rev. Acad. Colomb. Cienc. **XXVIII**, (107), 265 (2004)
2. Rodríguez, V.F., Jiménez, J.A., Adeva, P., Bohórquez, A., Pérez, G.A., Fernández, B.J., Chao, J.: Rev. Metal. Madrid **34**, 362 (1998)
3. Tjong, S.C., Swart, H.C.: Appl. Surf. Sci. **47**, 311 (1991)
4. Pérez, P., Pérez, F.J., Gómez, C., Adeva, P.: Corros. Sci. **44**, 113 (2002)
5. Cruz, B., Pérez Alcázar, G.A., Aguilar, Y.: Hyperfine Interact. **4**, 115 (1999)
6. Pérez Alcázar, G.A., Galvao da Silva, E., Paduani, C.: Hyperfine. Interact. **66**, 221 (1991)
7. Jen, S.U., Yao, Y.D., Huang, P.M., Lee, C.C., Chang, S.C.: J. Appl. Phys. **67**, (9), 4835 (1990)
8. Brand, R.A.: Nucl. Instrum. Methods Phys. Res. B **28**, 417 (1987)
9. MAUD program, Material Analysis Using Diffraction by L. Lutterotti. Version 2.046 (21 July 2006). Avalaible from http://www.ing.unitn.it/~maud/
10. Askelan, D.R., Phulé, P.P.: Essentials of Materials Science and Engineering, p. 105. Thomson, Canada (2004)
11. Chakrabarti, D.J.: Metall. Trans. B **8**, 121 (1977)
12. Umebayashi, H., Ishikawa, Y.: J. Phys. Soc. Jpn. **21**(No. 7) (1966), July
13. Ishikawa, Y., Endoh, Y.: J. Phys. Soc. Jpn. **23**(No. 2) (1967), August
14. Hashimoto, T., Ishikawa, V.: J. Phys. Soc. Jpn. **23**(No. 2) (1967), August
15. Bruna, P., Pradell, T., Crespo, D., Garcia-Mateo, C., Bhadeshia, H.K.D.H.: In: Garcia, M., Marco, J.F., Plazaola, F. (eds.) Industrial Applications of Mössbauer Effect, pp. 338–343. American Institute of Physics (2005)
16. Blachoswi, A., Ruebenbauer, K., Jura, J., Bonarski, J.T., Baudin, T., Penelle, R.: Nukleonica **48**, (Supplement I), S9–S12 (2003)
17. Pérez Alcázar, G.A., Plascak, J.A., Galvão da Silva, E.: Phys. Rev. B **2816** (1988)

Hyperfine Interact (2007) 175:71–75
DOI 10.1007/s10751-008-9590-3

# Mössbauer investigation of maghemite-based glycolic acid nanocomposite

J. G. Santos · L. B. Silveira · A. C. Oliveira ·
V. K. Garg · B. M. Lacava · A. C. Tedesco · P. C. Morais

Published online: 26 March 2008
© Springer Science + Business Media B.V. 2008

**Abstract** Transmission electron microscopy, X-ray diffraction and Mössbauer spectroscopy were used in the characterization of a nanocomposite containing magnetic nanoparticles dispersed in a glycolic acid-based template. Maghemite nanoparticles were identified as the iron oxide phase dispersed in the polymeric template. From the low-temperature Mössbauer data the amount of the iron-based, non-magnetic material at the nanoparticle surface was estimated as roughly one monolayer in thickness.

**Keywords** Maghemite · Magnetic nanocomposite · Mössbauer spectroscopy · Glycolic acid

J. G. Santos (✉) · L. B. Silveira
Departamento de Ciências Exatas e da Natureza, Fundação Universidade Federal de Rondônia,
Ji-Paraná, RO 78961-970, Brazil
e-mail: judes@unir.br

L. B. Silveira
e-mail: luciene@unir.br

A. C. Oliveira · V. K. Garg · B. M. Lacava · P. C. Morais
Instituto de Física, Universidade de Brasília, Núcleo de Física Aplicada, Brasília, DF 70910900, Brazil

A. C. Oliveira
e-mail: aderbal@unb.br

V. K. Garg
e-mail: garg@unb.br

B. M. Lacava
e-mail: bruno@unb.br

P. C. Morais
e-mail: pcmor@unb.br

A. C. Tedesco
FFCLRP-USP, Universidade de São Paulo, Preto Ribeirão, SP 14040-903, Brazil
e-mail: atedesco@usp.br

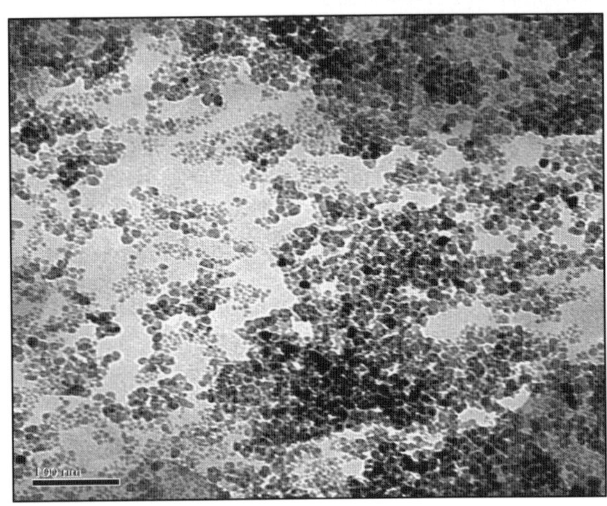


**Fig. 1** TEM micrograph of maghemite nanoparticles. The *horizontal bar* is 100 nm long

# 1 Introduction

In recent years the design and synthesis of biocompatible magnetic nanocomposites (BMNCs) has attracted intense interest, with special emphasis on their biological and biomedical applications, such as in cell sorting, contrast agents for magnetic resonance imaging, diagnosis, and cancer therapy [1, 2]. Much effort has been done to produce stable drug delivery systems using biocompatible magnetic nanocomposites. The engineering of these nanostructures, as for instance magnetic nanocapsules, magnetic nanoemulsions and magnetic fluids, demand a deep knowledge of both physical and chemical aspects [3–5]. In the present study glycolic acid (GA) microspheres were used as template to host maghemite nanoparticles (average diameter of 6.6 nm). The as produced BMNC was investigated using different techniques, including transmission electron microscopy (TEM), X-ray diffraction (XRD), and Mössbauer spectroscopy.

# 2 Experimental

Maghemite nanoparticles were chemically synthesized and peptized as a surfacted magnetic fluid (MF) sample, following the standard two-step procedure described in the literature [6]. In the first step, maghemite nanoparticles were synthesized by precipitating $Fe^{3+}$ aqueous ion in alkaline medium. In the second step oleic acid was used to peptize the as precipitated magnetic nanoparticles in organic medium at a final particle concentration of about $1.4 \times 10^{17}$ particle/cm$^3$. The average nanoparticle size and size dispersity were obtained from TEM micrographs using a Jeol JEM-1010 electron microscope. A drop of the MF sample containing about 0.00003% volume fraction was deposited on the sample holder (a 300 mesh copper grid covered with formvar) and dried at room temperature, in ambient air, before the TEM pictures were taken. Figure 1 shows a typical micrograph of spherical nanoparticles, from which a particle diameter histogram was obtained and curve-fitted using a log-normal distribution function.

Preparation of the GA-based nanocomposite containing magnetic nanoparticles was carried out using the solvent evaporation technique [7]. Before the evaporation procedure oleic acid-coated maghemite nanoparticles were mixed with glycolic acid solution (0.5 M).

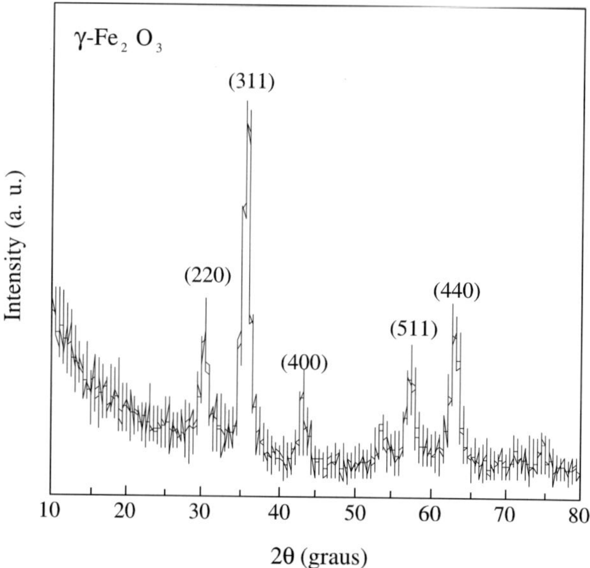

Fig. 2 X-ray diffractogram of the magnetic nanocomposite sample



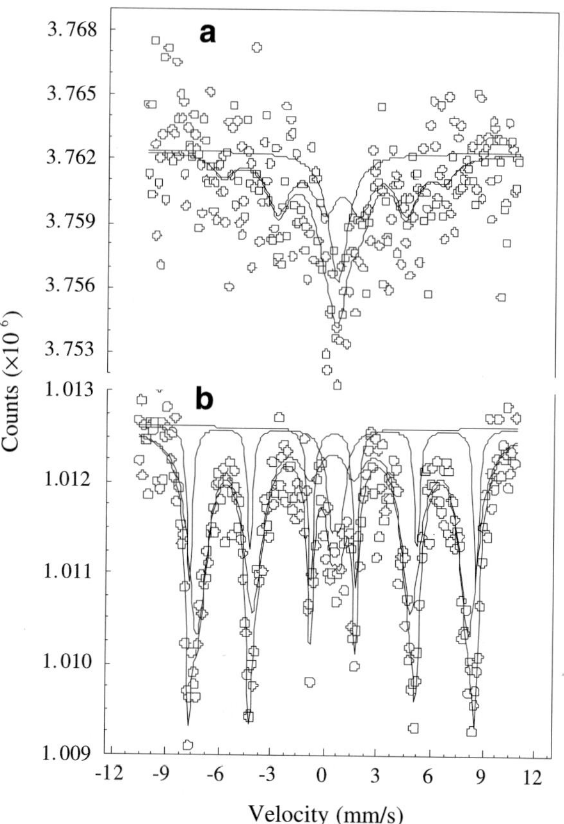

Fig. 3 Mössbauer spectra of the maghemite-based nanocomposite at **a** room temperature and at **b** liquid nitrogen temperature

**Table 1** Hyperfine Mössbauer parameters obtained from the fitting of the 77 and 300 K spectra of the magnetic nanocomposite sample

| Sample | T (K) | IS (mm/s) | B (kOe) | A (%) |
| --- | --- | --- | --- | --- |
| Composite | 77 | 0.60 | – | 7.4 |
| | | 0.42 | 479 | 21.0 |
| | | 0.45 | 506 | 71.5 |
| Composite | 300 | 0.37 | – | 30.0 |
| | | 0.33 | 396 | 70.0 |
| Bulk maghemite | 300 | 0.43 | 506 | – |

The hyperfine parameters of bulk maghemite taken from the literature [9] are included in the Table as well

X-ray diffraction measurements were recorded using a commercial Shimadzu (XRD-6000) system operating with the Cu-K$\alpha$ radiation source driven at 40 kV (1.5406 angstron line). Figure 2 shows a typical XRD spectrum of the maghemite-based nanocomposite sample.

Transmission Mössbauer spectra were recorded at 300 and 77 K, using a MCA (256 channels) and a Wissel constant acceleration transducer coupled to a 50 mCi $^{57}$Co/Rh source. Each plastic sample holder (1.7 cm diameter) contained about 80 mg of uniformly distributed and pressed sample. At 300 K the Mössbauer spectrum was fitted using one broad singlet and one sextet. However, at 77 K the Mössbauer spectrum was fitted using one doublet and two sextets. Figure 3a,b shows the Mössbauer spectra at 300 and 77 K, respectively. The hyperfine parameters obtained from the best fit of the Mössbauer spectra are shown in Table 1, namely the isomer shift (IS) and the internal field (B). Also, Table 1 shows the relative area (A) under the spectrum component. The IS shown in Table 1 was taken with respect to natural iron.

## 3 Results and discussion

The curve-fitting of a 300-count particle diameter histogram allowed us to obtain the average nanoparticle diameter (6.6 nm) and the diameter standard deviation (0.15). Such a standard deviation value characterizes the prepared sample as containing monodisperse nanoparticles. Note that nanoparticle size distributions with diameter standard deviation above 0.20 is considered as polydisperse [8].

The X-ray diffraction of the magnetic material incorporated in the nanocomposite sample (see Fig. 2) revealed peaks we labelled as (220), (311), (400), (511) and (440), whose positions and relative intensities are in excellent agreement with the ASTM data for maghemite (International Center for Diffraction Data, 2000).

Table 1 shows the hyperfine parameters obtained from the best curve-fitting of the Mössbauer spectra of the nanocomposite sample. Also, Table 1 shows the hyperfine parameters of bulk maghemite obtained from the literature [9]. The doublet observed at 300 K is claimed to be related to the non-magnetic nanoparticle surface layer. The 30% associated to the area under the doublet (300 K) represents a surface layer of about 0.36 nm in thickness, roughly representing one atom thickness in average. This information is consistent with the well-know idea of the so-called magnetic dead layer associated to nanosized magnetic particles [10]. Nevertheless, at 77 K, the observed value of 7.4% (see Table 1) associated to the area under the broad Mössbauer singlet would represent a non-magnetic surface layer of about 0.08 nm in thickness. Though average in size the 0.08 nm

value we found from the fitting of the 77 K Mössbauer spectrum is thinner than just one atom in thickness. Considering the monodisperse characteristic of the synthesized nanoparticles this finding may indicate that the broad singlet observed at 77 K could be due to an iron-based material not related to maghemite. The possibility of having a material different from maghemite is consistent with the oversized thickness (0.36 nm) we found for the dead magnetic layer at 300 K.

## 4 Conclusions

From the transmission electron microscopy micrographs the synthesized nanoparticles incorporated in the glycolic acid-based template are spherically shaped with a monodisperse size profile, presenting an average diameter and diameter dispersion of about 6.6 nm and 0.15, respectively. The X-ray data indicated that maghemite nanoparticles were synthesized and incorporated in the polymeric template. Mössbauer spectroscopy supports the X-ray data in providing evidence for the synthesis and incorporation of maghemite nanoparticles in the polymeric template. In addition, room-temperature Mössbauer data indicated that a non-magnetic surface layer of about one atom thick was identified around the maghemite nanoparticles.

**Acknowledgements** The authors acknowledge the financial support of the Brazilian agencies, DCEN/PROPLAN-UNIR, CNPq, CAPES, and FINATEC.

## References

1. Berry, C.C., Curtis, A.S.G.: J. Phys. D: Appl. Phys. **36**, R198 (2003)
2. Jordan, A., Wust, P., Scholz, R., Faehling, H., Felix, R.: In: Hafeli, U., Schutt, W., Teller, J., Zborowski, M. (eds.) Scientific and clinical applications on magnetic carriers, p. 569. Plenum, New York (1997)
3. Goetze, T., Gansau, C., Buske, N., Roeder, M., Gornert, P., Bahr, M.: J. Magn. Magn. Mater. **252**, 399 (2002)
4. Simioni, A.R., Martins, O.P., Lacava, Z.G.M., Azevedo, R.B., Lima, E.C.D., Lacava, B.M., Morais, P.C., Tedesco, A.C.J.: Nanosci. Nanotechnol. **6**, 2413 (2006)
5. Macaroff, P.P., Primo, F.L., Azevedo, R.B., Lacava, Z.G.M., Morais, P.C., Tedesco, A.C.: IEEE Trans. Magn. **42**, 3596 (2006)
6. Morais, P.C., Santos, R.L.A., Pimenta, C.M., Azevedo, R.B., Lima, E.C.D.: Thin Sol. Films **515**, 266 (2006)
7. Avivi, S., Felner, I., Novik, I., Gedanken, A.: Biochem. Biophys. Acta **1527**, 123 (2001)
8. Lacava, B.M., Azevedo, R.B., Silva, L.P., Lacava, Z.G.M., Skeff Neto, K., Buske, N., Bakuzis, A.F., Morais, P.C.: Appl. Phys. Lett. **77**, 1876 (2000)
9. Kuzmann, E., Nagy, S., Vértes, A., Weiszburg, T.G., Garg, V.K.: In: Vértes, A., Nagy, S., Süvegh, K. (eds.) Geological and mineralogical applications of Mössbauer spectroscopy in Nuclear methods in mineralogy and geology: techniques and applications, p. 285. Plenum, New York (1998)
10. Di Marco, M., Port, M., Couvreur, P., Dubernet, C., Ballirano, P., Sadun, C.: J. Amer. Chem. Soc. **128**, 10054 (2006)

Hyperfine Interact (2007) 175:77–84
DOI 10.1007/s10751-008-9591-2

# Structural, Mössbauer and magnetic studies on Mn-substituted barium hexaferrites prepared by high energy ball milling

Puneet Sharma · R. A. Rocha · S. N. de Medeiros ·
A. Paesano Jr · B. Hallouche

Published online: 9 April 2008
© Springer Science + Business Media B.V. 2008

**Abstract** In the present study, $BaFe_{12-x}Mn_xO_{19}$ hexaferrites were prepared by high-energy ball milling and subsequent thermal annealing. The structural and magnetic characterizations were carried out by X-ray diffraction, Mössbauer spectroscopy and magnetization measurements. The analyses showed that manganese occupies all iron sites, decreasing the magnetization and increasing the coercivity.

**Keywords** Barium hexaferrite · High-energy ball milling · Mössbauer spectroscopy

## 1 Introduction

Since the discovery of the M-type hexagonal ferrites in 1950s, it has being of great interest due to its application as permanent magnetic materials and perpendicular recording media [1, 2]. The main reason for its great success is its low cost at moderate magnetic properties. Many attempts have been made to improve its magnetic properties by divalent-tetravalent and trivalent cationic substitution for $Fe^{3+}$ ion situated at the five different crystallographic sites, i.e., the octahedral (12k, $4f_2$, 2a), tetrahedral ($4f_1$) and trigonal bipyramidal (2b) sites. The most extensively studied divalent-tetravalent substitutions were Co–Ti [3], Co–Sn [4], Ni–Ti [5], Zn–Ti [6, 7], Zn–Zr [8], whereas some studied trivalent cations are Al, Ga, Cr, Sc and In [9–11]. In addition, some trivalent metal cations that can also exist in divalent and tetravalent state such as manganese, have been investigated. Collomb et al. [12], *e.g.*, carried out a magnetic structure study on manganese substitution, considering its $Mn^{2+}$,

P. Sharma · R. A. Rocha · S. N. de Medeiros · A. Paesano Jr (✉)
Departamento de Física, Universidade Estadual de Maringá, Av. Colombo, 5790, 87.020-900 Maringá,
Paraná, Brazil
e-mail: paesano@wnet.com.br

B. Hallouche
Departamento de Química e Física, UNISC, Santa Cruz do Sul, Rio Grande do Sul, Brazil

$Mn^{3+}$ and $Mn^{4+}$ states at the different crystallographic sites. Turrili et al. [13] also investigated $Mn^{2+}$-$Ti^{4+}$ substitution. Recently, electron energy-loss spectroscopy (EELS) studies on some Mn-ferrites and Mn-substituted barium hexaferrites suggested that manganese remains in $Mn^{3+}$ ionic state [14].

In the present study, Mn-substituted barium hexaferrites were prepared by high-energy ball milling. The influence of $Mn^{3+}$ ion substitution on the structure and magnetic properties was systematically investigated and discussed.

## 2 Experimental

$BaFe_{12-x}Mn_xO_{19}$ samples with $X=0.5$, 1.0, 1.5, and 2.0 were prepared by high-energy ball milling for 30 h in free atmosphere, using laboratory grade compounds $BaCO_3$, $Fe_2O_3$, and $Mn_2O_3$, in a planetary mill with hardened steel vial and balls. The ball-to-powder ratio was 10:1 and the speed rotation was 300 rpm. Further, the milled powders were annealed at 1050°C for 2 h, also in free atmosphere, using a resistance furnace. The heating and cooling rates were 51°C/min.

The phase characterization of powders was carried out in a SHIMADZU-6000 X-ray diffractometer. The XRD patterns of heat-treated samples were refined by the Rietveld method (Fullprof Suite-2000 software). Mössbauer characterizations were performed at room temperature and 80 K, in the transmission geometry, using a conventional spectrometer operating in a constant acceleration mode. The gamma rays were provided by a $^{57}Co(Rh)$ source. The Mössbauer spectra were analyzed with a non-linear least-square routine, with Lorentzian line shape. The magnetic properties of randomly pressed powders were measured in a vibrating sample magnetometer, with a maximum field of 1.5 T.

## 3 Results and discussions

The X-ray diffractograms of the non-substituted and $X=2.0$ substituted barium hexaferrite samples are shown in Fig. 1. Both samples show the barium hexaferrite as a single phase. However, a minor peak of hematite is also observed in the non-substituted sample. No peaks for the $Mn_2O_3$ phase were observed. Also, the presence of metallic iron was not detected in the diffractograms, which was further confirmed by Mössbauer spectroscopy. Thus, no significant contamination from the milling media is expected.

Figure 2 shows the lattice parameters, $a$ and $c$, as a function of the Mn content, as obtained by Rietveld analysis. The decrease in the lattice parameter $c$ means that the unit cell shrinks perpendicularly to the basal plane.

Figure 3 shows the Mössbauer spectra for the $X=0.0$, 1.0, 2.0 samples, taken at room temperature (Fig. 2a) and 80 K (Fig. 2b). They were fitted with five and six discrete sextets, respectively. For the non-substituted sample, the fit was carried out constraining the subspectral areas to the ratio 6:2:2:1:1 for the 12k, $4f_1$, $4f_2$, 2a and 2b sites, respectively. With the purpose of simplification, no component for the hematite phase was considered, due to its minor occurrence. However, for substituted samples the subspectral areas of all five sites were kept free. The fitted parameters were the hyperfine magnetic fields ($B_{hf}$), the quadrupolar shifts (QS), the isomer shifts (IS) and the line widths ($\Gamma$) besides the subspectral areas.

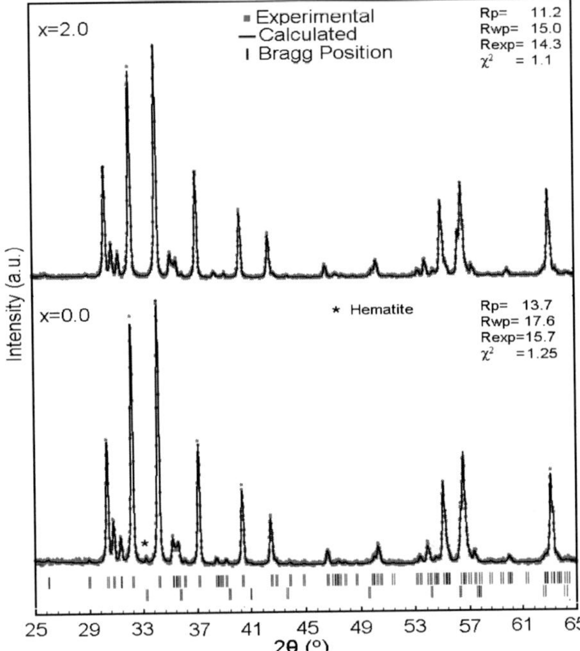

**Fig. 1** X-ray diffraction patterns for the $BaFe_{12}O_{19}$ and $BaFe_{10}Mn_2O_{19}$ samples



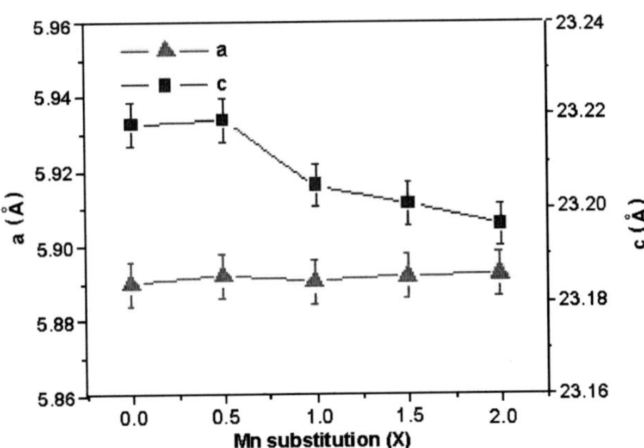

**Fig. 2** Variation in the lattice parameter with the Mn content



The fit for the $X \geq 1$ samples measured at RT required six sextets. Because of the IS and QS similarity and after inspecting the relative areas, we concluded that the 12 K was split in two components, henceforth designed as $12k_1$ and $12k_2$. The split may be attributed to different neighborhoods for the iron in the crystallographic site 12k, resultant from the manganese substitution. One of these new components has a hyperfine field weakened relatively to the other, indicating a little smaller magnetic moment for the iron in this specific neighborhood. On the other hand, a small difference between the quadrupolar splittings of the two components can also be verified.

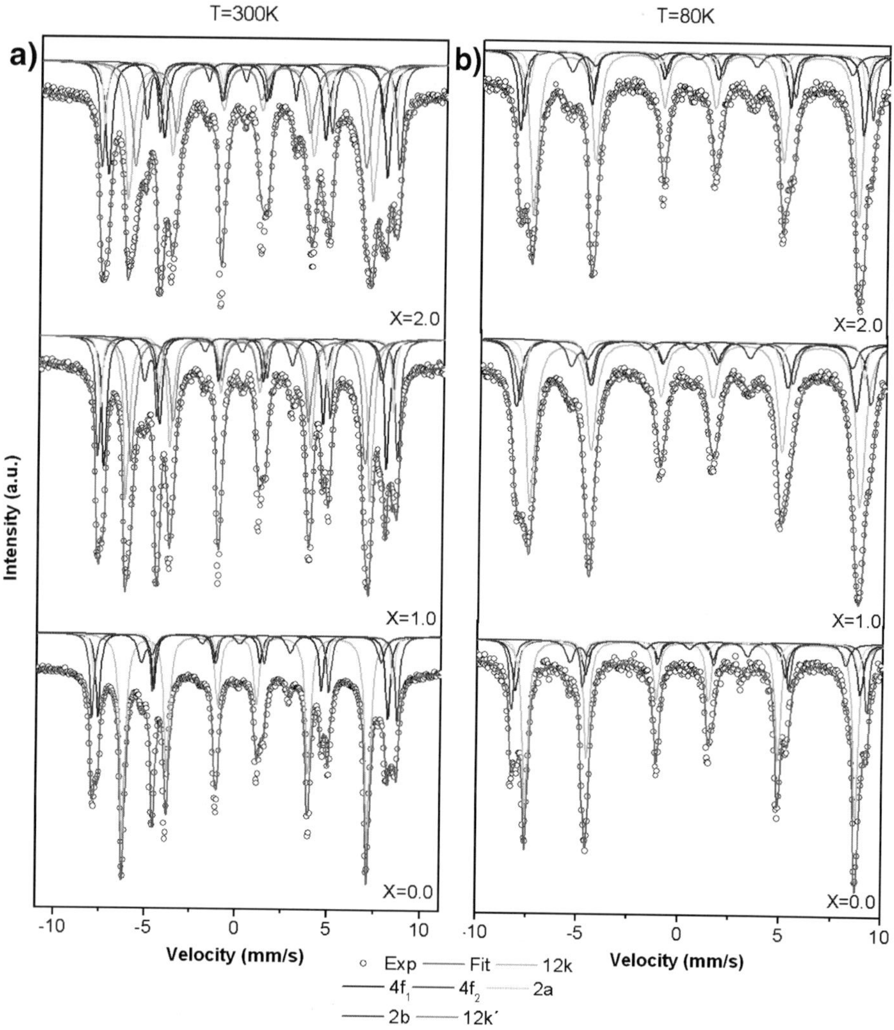

Fig. 3 Mössbauer spectra taken at 300 K (**a**) and 80 K (**b**) for Mn-substituted ($X$=0.0, 1.0 and 2.0) barium hexaferrites

Interestingly, the split relative to the 12k site disappears as the temperature goes down since, again, five components were enough for the fits. This reveals that both iron magnetic moments are temperature sensible in a diverse manner, though reaching virtually the same value at 80 K. The hyperfine parameters from measurements at 300 K and 80 K are given in Tables 1 and 2, respectively. It can be seen that the $B_{hf}$ values decrease monotonically with the manganese content, including the average values for the 12k's components, revealing the progressive weakening of the magnetic moments of iron at all sites. On the contrary, reducing the temperature, the magnetic hyperfine fields increase up to 20% of their RT values.

✎ Springer

**Table 1** Hyperfine parameters and subspectral areas for the $BaFe_{12-X}Mn_XO_{19}$ samples measured at 300 K

| Mn content ($X$) | Site | $B_{hf}$ ($T$) | IS[a] (mm/s) | QS (mm/s) | $\Gamma$ (mm/s) | $S$ (%) |
|---|---|---|---|---|---|---|
| 0.0 | 12k | 41.4 | 0.24 | 0.40 | 0.35 | 50.0 |
|  | 4f$_1$ | 49.0 | 0.25 | 0.25 | 0.29 | 16.7 |
|  | 4f$_2$ | 51.6 | 0.27 | 0.16 | 0.28 | 16.7 |
|  | 2a | 50.6 | 0.25 | 0.09 | 0.25 | 8.3 |
|  | 2b | 40.5 | 0.21 | 2.15 | 0.46 | 8.3 |
| 0.5 | 12k | 41.0 | 0.24 | 0.40 | 0.39 | 50.4 |
|  | 4f$_1$ | 48.3 | 0.16 | 0.20 | 0.28 | 17.2 |
|  | 4f$_2$ | 51.1 | 0.26 | 0.17 | 0.26 | 16.4 |
|  | 2a | 49.9 | 0.26 | 0.09 | 0.24 | 7.8 |
|  | 2b | 40.2 | 0.18 | 2.13 | 0.45 | 8.2 |
| 1.0 | 12k$_1$ | 41.2 | 0.24 | 0.41 | 0.37 | 25.8 |
|  | 12k$_2$ | 39.5 | 0.25 | 0.40 | 0.38 | 24.5 |
|  | 4f$_1$ | 47.7 | 0.17 | 0.18 | 0.32 | 17.7 |
|  | 4f$_2$ | 50.6 | 0.28 | 0.19 | 0.32 | 16.3 |
|  | 2a | 49.5 | 0.25 | 0.13 | 0.25 | 7.8 |
|  | 2b | 40.0 | 0.17 | 2.10 | 0.45 | 7.9 |
| 1.5 | 12k$_1$ | 41.2 | 0.24 | 0.41 | 0.45 | 28.2 |
|  | 12k$_2$ | 39.2 | 0.24 | 0.38 | 0.44 | 23.4 |
|  | 4f$_1$ | 47.4 | 0.18 | 0.17 | 0.39 | 18.4 |
|  | 4f$_2$ | 50.4 | 0.28 | 0.20 | 0.33 | 15.8 |
|  | 2a | 49.1 | 0.27 | 0.16 | 0.27 | 7.0 |
|  | 2b | 40.0 | 0.17 | 2.03 | 0.40 | 7.2 |
| 2.0 | 12k$_1$ | 41.2 | 0.24 | 0.39 | 0.50 | 27.9 |
|  | 12k$_2$ | 38.9 | 0.24 | 0.39 | 0.57 | 23.9 |
|  | 4f$_1$ | 47.0 | 0.19 | 0.14 | 0.41 | 19.0 |
|  | 4f$_2$ | 50.1 | 0.29 | 0.22 | 0.35 | 15.0 |
|  | 2a | 48.9 | 0.27 | 0.18 | 0.26 | 7.0 |
|  | 2b | 39.7 | 0.17 | 2.0 | 0.44 | 7.2 |

[a] Relative to $\alpha$–Fe

**Table 2** Hyperfine parameters and subspectral areas for the $X$=0.0, 1.0, 2.0 samples measured at 80 K

| Mn content ($X$) | Site | $B_{hf}$ ($T$) | IS[a] (mm/s) | QS (mm/s) | $\Gamma$ (mm/s) | $S$ (%) |
|---|---|---|---|---|---|---|
| 0.0 | 12k | 50.6 | 0.35 | 0.40 | 0.35 | 50.0 |
|  | 4f$_1$ | 52.9 | 0.36 | 0.10 | 0.40 | 16.7 |
|  | 4f$_2$ | 54.3 | 0.37 | 0.23 | 0.30 | 16.7 |
|  | 2a | 51.8 | 0.28 | 0.24 | 0.21 | 8.3 |
|  | 2b | 41.8 | 0.27 | 2.20 | 0.43 | 8.3 |
| 1.0 | 12k | 50.4 | 0.38 | 0.42 | 0.61 | 50.8 |
|  | 4f$_1$ | 51.9 | 0.20 | 0.0 | 0.48 | 18.5 |
|  | 4f$_2$ | 53.6 | 0.47 | 0.19 | 0.42 | 16.6 |
|  | 2a | 51.8 | 0.22 | 0.57 | 0.22 | 6.3 |
|  | 2b | 42.9 | 0.32 | 2.10 | 0.55 | 7.8 |
| 2.0 | 12k | 49.6 | 0.34 | 0.33 | 0.74 | 51.6 |
|  | 4f$_1$ | 52.4 | 0.24 | 0.0 | 0.35 | 18.9 |
|  | 4f$_2$ | 53.1 | 0.48 | 0.20 | 0.34 | 15.3 |
|  | 2a | 51.4 | 0.28 | 0.50 | 0.28 | 6.9 |
|  | 2b | 42.7 | 0.35 | 1.75 | 0.61 | 7.3 |

[a] Relative to $\alpha$–Fe

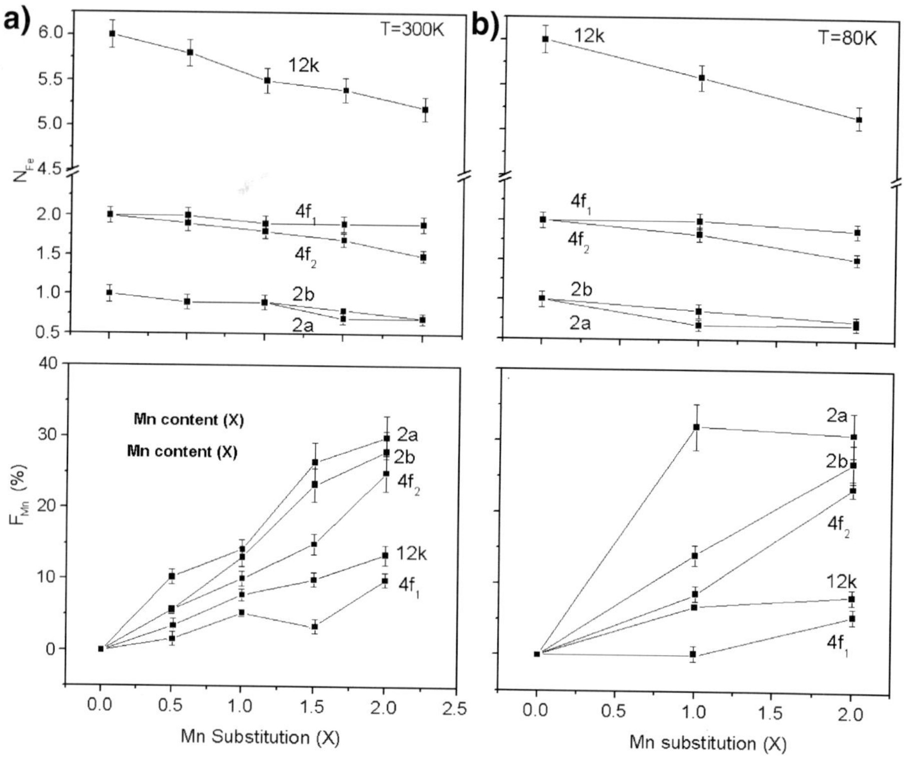

**Fig. 4** Site occupation number for iron ($N_{Fe}$) and site occupation of $Mn^{3+}$ ions on each site at 300 K (**a**) and 80 K (**b**)

In order to estimate the $Mn^{3+}$ ion occupation, the site occupation numbers for iron ion, $N_{Fe}$, and the associated $Mn^{3+}$ ion occupancy fraction, $F_{Mn}$, were calculated by using the formulae [8]

$$N_{Fe}(i) = C_{Fe} \left[ \frac{S(i)}{\sum\limits_{i=1}^{5} S(i)} \right]$$

$$F_{Mn} = \frac{N(i) - N_{Fe}(i)}{N(i)} \times 100\%,$$

where $C_{Fe}$ denotes the iron content (i.e., $C_{Fe}=12-X$), $S(i)$ is the subspectral area of the $i$th site obtained from the least square fit and $N(i)$ is the occupation number for the $i$th site.

It can be seen from the Fig. 4 (a and b), that $Mn^{3+}$ ions occupy all the sites, although the $4f_1$ site is least substituted which is diverse to previously reported studies [12]. The site occupation numbers and the site occupation fractions, as obtained from the Mössbauer characterizations at RT and 80K, reveal satisfactory consistency, exception to the 2a site that, for $X=1.0$, shows an occupation fraction for $T=80$ K almost three times the value obtained for RT. In our opinion, this is much more a product of the limited resolution (i.e.,

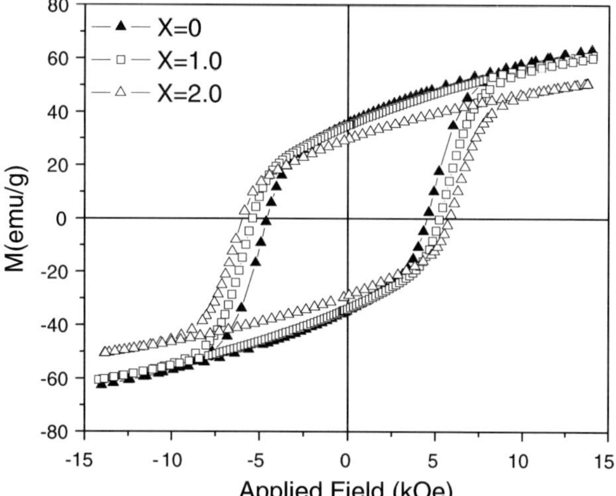

**Fig. 5** RT magnetization curves, as a function of the applied field, for the $BaFe_{12}O_{19}$, $BaFe_{11}Mn_{1.0}O_{19}$ and $BaFe_{10}Mn_2O_{19}$ samples

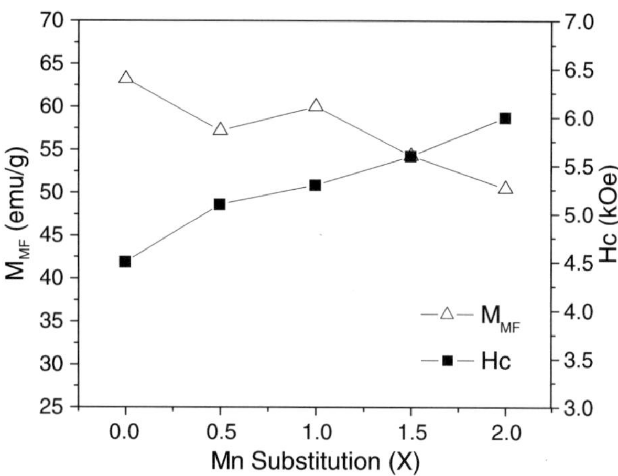

**Fig. 6** Magnetization at maximum field and coercivity, as a function of Mn content, for the $BaFe_{12-X}Mn_XO_{19}$ samples

the uncertainty) of the fitted area parameters than a differentiated change with temperature in the f factors respective to the various iron sites in the hexaferrite structure.

Figure 5 shows the RT magnetization curves, as a function of the applied field, for the $X=$ 0.0, 1.0 and 2.0 samples. A lower magnetization at maximum field, $M_{MF}$, and a higher coercivity are evident in the Mn-substituted samples.

The variation of $M_{MF}$ with the Mn substitution is shown in the Fig. 6. The lower saturation magnetization could be understood by the site occupation of $Mn^{3+}$ ions, which takes place for all the sites, spin up (12k, 2a, 2b) or spin down ($4f_1$, $4f_2$) sites. However, the percent site occupation in the spin up sites is quite higher as compared to the spin down sites. This higher level of occupation at spin up sites may result in a smaller magnetic moment decreasing, therefore, the $M_{MF}$ in Mn-substituted samples.

Figure 6 also shows the variation of coercivity upon Mn substitution. The coercivity increases with the substitution amount. This may attributed to the decrease in the hexagonal

 Springer

$c$ axis, since the shrinkage would lower the $Fe^{3+}$–O–$Fe^{3+}$ bond lengths, enhancing the superexchange interaction and, consequently, the coercivity. Our previous investigation revealed that lower substitution levels do not affect the lattice parameters. The observed increase in coercivity was mainly ascribed to the reduction in the grain size [15].

## 4 Conclusions

Mn-substituted barium hexaferrites were prepared by high-energy ball milling and subsequent heat treatment at 1050°C. It was observed from the Mössbauer investigation that Mn substitution above $X=0.5$ affects the 12k site, splitting it in two, $12k_1$ and $12k_2$. Both are further transformed to only one (12k) site at 80 K. The occupation number and occupancy fraction of Mn ion suggest that $Mn^{3+}$ ion substituted all five crystallographic sites. The magnetization decreases with the substitution amount due to the dilution of the magnetic structure. The increase in coercivity is due to the decrease in lattice parameter, $c$, which may enhance the superexchange interaction between neighboring $Fe^{3+}$ ions.

**Acknowledgements** The authors would like to thank to the Brazillian agencies, CNPq (Bolsa PDJ) and CAPES, for financial support.

## References


1. Speliotis, D.E.: IEEE Trans. Magn. **23**, (5), 3143 (1987)
2. Kojima, H.: Ferromagnetic materials. In: Wohlfarth, E. P. (ed.), vol. 3, Chapter 5 (1982)
3. An, S.Y., Shim, I., Kim, C.S.: J. Appl. Phys. **91**, (10), 8465 (2002)
4. Zhou, X.Z., Morrish, H.: J. Appl. Phys. **75**, (10), 5556 (1994)
5. Turrili, G., Licci, F., Paoluzi, A., Besagni, T.: IEEE Trans. Magn. **24**, (4), 2146 (1987)
6. Wartewig, P., Krause, M.K., Esquinazi, P., Rösler, S., Sonntag, R.: J. Magn. Magn. Mater. **192**(1), 83 (1999)
7. Gonzalez-Angeles, A., Suarez, G.M., Gruskova, A., Papanova, M., Slama, J.: Mater. Lett. **59**, 26 (2005)
8. Li, Z.W., Ong, C.K., Yang, Z., Wei, F.W., Zhou, X.Z., Zhao, J.H., Morrish, A.H.: Phys. Rev. B **62**, (10), 6530 (2000)
9. Wang, S., Ding, J., Shi, Y., Chen, J.: J. Magn. Magn. Mater. **219**, 206 (2000)
10. Clark, T.M., Evans, B.J., Thompson, G.K.: J. Appl. Phys. **85**, (2), 5229 (1999)
11. Ounnunkad, S., Winotai, P.: J. Magn. Magn. Mater. **301**, (2), 292 (2006)
12. Collomb, A., Obradors, X., Isalgué, A., Fruchart, D.: J. Magn. Magn. Mater. **69**, 317 (1987)
13. Turilli, G., Licci, F., Rinaldi, S.: J. Magn. Magn. Mater. **59**, 127 (1986)
14. Schmid, H.K., Mader, W.: Micron. **37**, 426(2006)
15. Sharma, P., Rocha, R.A., Medeiros, S.N., Hallouche, B., Paesano Jr., A.: J. Magn. Magn. Mater. **316**, 29 (2007)

Hyperfine Interact (2007) 175:85–90
DOI 10.1007/s10751-008-9592-1

# Provenance study of obsidians from the archaeological site of La Maná (Ecuador) by electron spin resonance (ESR), SQUID magnetometry and $^{57}$Fe Mössbauer spectroscopy

M. Duttine · R. B. Scorzelli · G. Poupeau ·
A. Bustamante · A. V. Bellido · R. M. Lattini ·
N. Guillaume-Gentil

Published online: 6 May 2008


**Abstract** Obsidians from major Ecuadorian sources (outcrops) were analyzed by electron spin resonance, SQUID magnetometry and $^{57}$Fe Mössbauer spectroscopy. If the last technique allows to discriminate obsidians from the Quiscatola source, an association of ESR with SQUID magnetometry permits to differentiate obsidians from the sources of Cotopaxi volcano, from the Quiscatola and Mullumica-Callejones sources of the Chacana caldera and to infer that the 12 analyzed obsidians from the pre-Hispanic site of La Maná come from the Mullumica-Callejones source.

**Keywords** Obsidian · Electron spin resonance · Magnetic properties ·
$^{57}$Fe Mössbauer spectroscopy · Obsidian provenance · Ecuador

M. Duttine · R. B. Scorzelli (✉)
Centro Brasileiro de Pesquisas Físicas, Rio de Janeiro, Brazil
e-mail: scorza@cbpf.br

M. Duttine
e-mail: m.duttine@yahoo.fr

G. Poupeau
Institut de Recherche sur les Archéomatériaux, UMR 5060 CNRS–Université Bordeaux 3, Pessac,
France

A. Bustamante
Universidade Nacional Mayor de San Marco, Lima, Peru

A. V. Bellido · R. M. Lattini
Departamento de Físico-Química, UFF, Instituto de Química, Rio de Janeiro, Brazil

N. Guillaume-Gentil
Institut de Préhistoire et des sciences de l'Antiquité, Université de Neuchâtel, Neuchâtel, Switzerland

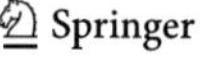

## 1 Introduction

Obsidian is a dark volcanic glass ($SiO_2 > 65\%$) used by prehistoric populations to produce various kinds of tools, arrowheads, ritual objects, etc. The 'sources' (outcrops) of raw material are often situated far away (up to several hundreds of kilometers) from the archaeological sites where obsidian artifacts are discovered. Thenceforwards the determination of the geographical origin of 'archaeological' obsidians provides valuable information about the selection of lithic resources by ancient human groups and eventually about the existence of communication routes and exchange networks. This kind of 'provenance' study is based on the comparison of specific properties of archaeological vs. geological samples collected in potential sources (i.e. reachable at the considered archaeological periods). These comparisons may bear on the elemental compositions, the formation ages and/or the structural properties of obsidians.

In Ecuador, earlier obsidian provenance studies were based on elemental composition [1] fission track (FT) dating [2] or both [3, 4]. We introduce here a new approach, using electron spin resonance (ESR), SQUID magnetometry and [57]Fe Mössbauer spectroscopy ([57]Fe-MS), which already proved their capabilities for obsidian provenance studies in the Western Mediterranean area [5–7]. This first test-program compares archaeological samples from the pre-Hispanic site of La Maná [8] to source samples whose FT age and/or elemental composition were previously determined [2–4, 9, 10].

## 2 Sampling and experimental conditions

The analyzed obsidians come mostly from the Mullumica (six samples, CM and CSM) and Callejones or Oyacachi (four samples, OYA) lava flows, from the Quiscatola or Yanaurcu sources (two samples, QSC) and from the river bed of the Rio Guambi (two samples, GMB). The Rio Guambi source is a secondary deposit, where obsidians eroded from the Mullumica flow are found as rounded pebbles. The Mullumica-Callejones lava flows and the Quiscatola-Yanaurcu volcanic breccias, localized in the Sierra de Guamani (Eastern Andean Cordillera, some 30 km E of Quito, Ecuador) belong to a major volcanic structure, the Chacana caldera [10]. Six other samples (CTX) come from various pyroclastic units associated to the Cotopaxi volcano, 50 km SE of Quito. The archaeological samples come from the *tola* (artificial mound) site of La Maná, localized between the Ecuadorian cities of La Cadena and Quevedo (about 140 km SW of Quito). This site was occupied from the Early Formative (3,600–1,800 B.C.) until the Spanish conquest (sixteenth century). In Ecuador, nearly all archaeological obsidians come from the sources of the Sierra de Guamani [1–4].

ESR spectra were recorded in X-band frequency (9.75 GHz) with a commercial spectrometer (Brüker). The main experimental parameters were: microwave power 10 mW and modulation amplitude 8.0 G at 100 kHz. The magnetization curves ($M$ vs. $H$ cycles where $M$ is the magnetization and $H$ the applied field) were obtained under a maximum applied field of 1.2 T using a commercial SQUID magnetometer (Quantum Design). Mössbauer transmission spectra were collected with conventional spectrometer using a standard $\gamma$-source [57]Co/Rh (activity 50 mCi). The spectra were fitted using the NORMOS program. All measurements were performed at room temperature.

 Springer

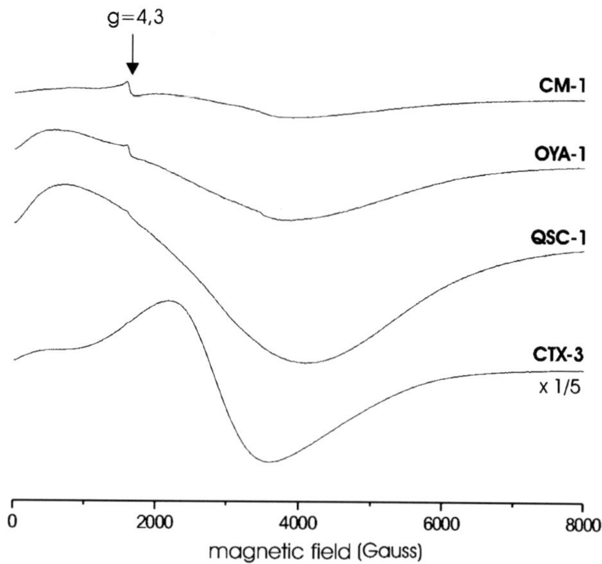


**Fig. 1** Examples of X-band room temperature ESR spectra of Ecuadorian obsidians

## 3 Results and discussion

The X-band ESR spectra of the geological samples are mainly composed of three different signals: a broad (peak-to-peak line width $\Delta H_{pp}$ ranging from 2,000 to 3,200 G) and intense resonance line due to micro-crystallites of mixed iron oxides (hematite, magnetite and related spinels), a narrower signal (20 G$\leq\Delta H_{pp}\leq$175 G) at $g_{eff}\approx$4.3 with a shoulder at about $g_{eff}\approx$9.8 associated to isolated $Fe^{3+}$ ions in the glassy matrix and another line at $g\approx$2 ascribed to $Fe^{3+}$ condensed clusters [6] (Fig. 1). For each sample the peak-to-peak amplitude ($I_{pp}$) and line width ($\Delta H_{pp}$) of these three components were measured using combined simulation and fitting procedures. As a preliminary result, the detection and quantification of the $g$=4.3 line allow one to separate the Cotopaxi obsidians from the others.

From the magnetization experiments (cycles $M$ vs. $H$), several parameters were measured: the magnetization at saturation ($M_S$), the remanent magnetization ($M_R$), the coercive field ($H_C$) and the susceptibility ($\chi$). Significant variations of the magnetization at saturation, from 0.009 to 0.470 emu/g and of the coercive field, ranging from 95 to 210 Oe, were observed among all the samples analyzed. These two parameters separate clearly the Quiscatola obsidians from the other Ecuadorian obsidians.

The $^{57}$Fe-MS spectra of geological samples are mainly composed of doublets, associated to $Fe^{3+}$ and $Fe^{2+}$ in two different sites and, for some samples, sextets attributed to iron oxides (hematite and/or magnetite) (Fig. 2). Multivariate statistical analysis using cluster analysis methods in an $n$-dimensions space was used to differentiate the Ecuadorian obsidians from Mössbauer hyperfine parameters, the $Fe^{2+}/Fe^{3+}$ ratio and the relative contribution of the magnetic components in the MS spectra. Only Quiscatola obsidians can be clearly distinguished from the other samples.

It is therefore possible to distinguish the Mullumica-Callejones, the Quiscatola and the Cotopaxi obsidians if an ESR–$^{57}$Fe-MS or an ESR–SQUID magnetometry coupling is

Fig. 2 Typical room temperature Mössbauer spectra for Ecuadorian obsidians. The *solid lines* show the results of the fitting procedure

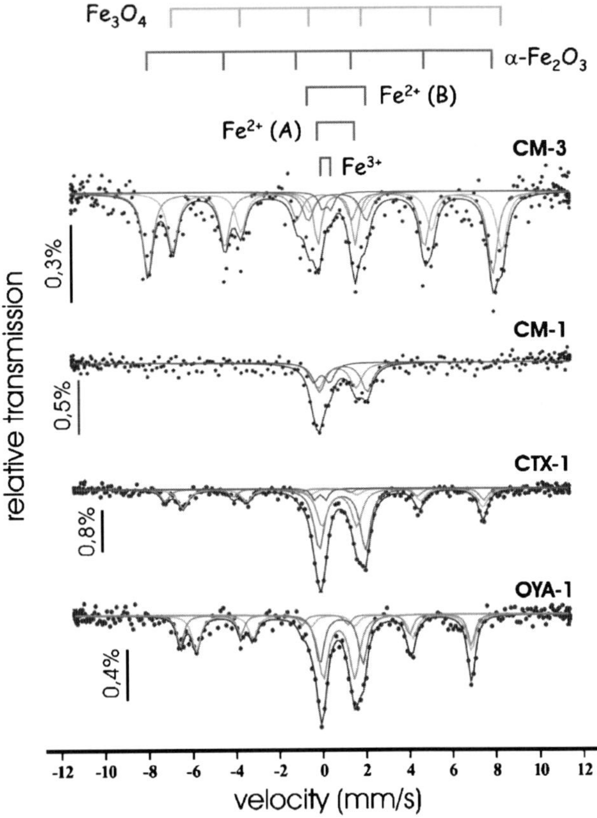


Fig. 3 3D diagram involving ESR (peak-to-peak intensity of the $g$=4.3 line, $I_{ESR}$, and its line width, $\Delta H_{ESR}$) and magnetization parameters (coercive field, $H_C$) for Ecuadorian obsidians. The La Maná archaeological samples are represented by *gray spheres*

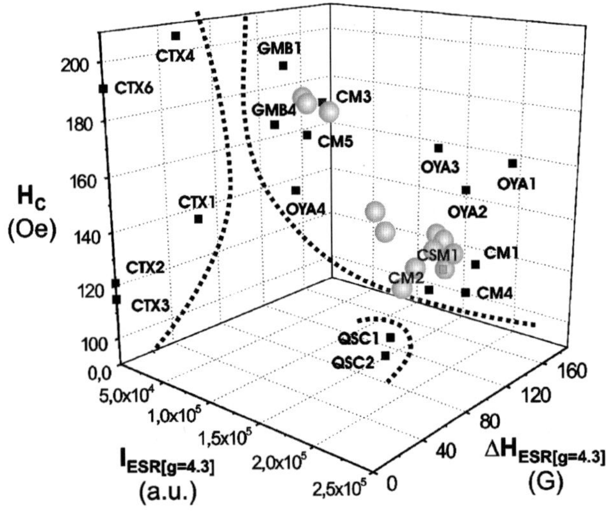

used. In the later case several 3D diagrams can discriminate obsidians of these sources. As an example, the Fig. 3 show that it is possible, using two ESR parameters $I_{\mathrm{ESR}[g=4.3]}$, $\Delta H_{\mathrm{ESR}[g=4.3]}$ and $H_{\mathrm{C}}$, to differentiate the Cotopaxi, the Quiscatola and the Mullumica-Callejones-Guambi obsidians. The dispersion of data points for the later might be due at least partly to the fact that the Mullumica and Callejones flows result from the incomplete mixing of two parent-magmas shortly before the eruptive episodes as demonstrated by their elemental composition [10].

Twelve archaeological samples from La Maná were analyzed by ESR and by SQUID magnetometry. All these obsidians can be attributed to the Mullumica-Callejones source (Fig. 3), in agreement with their elemental composition [11].

## 4 Conclusion

ESR and SQUID magnetometry appear as promising techniques for obsidian provenance studies in the Ecuadorial Andean belt. These methods are particularly attractive for their low cost, simple sample preparations (powders) and short data acquisition times. Moreover, their coupled use need only an aliquot of <50 mg. Mössbauer spectroscopy of $^{57}$Fe may also contribute to source discrimination, but its need of a 250-mg powder and a long data acquisition time will limit its use as a complement in doubtful cases.

An extended on-going program on obsidians from other (Peruvian to Colombian) Andean sources is in progress, which will allow us to more fully evaluate the potentials and limits of these techniques in provenance studies.

**Acknowledgements** This work was supported by the Centro Brasileiro de Pesquisas Físicas, Brazil. M.D. is indebted to the CNPq for financial support during his stay at CBPF.

## References

1. Asaro, F., Salazar, H.V., Michel, R.L., Burger, R.L., Stross, F.H.: Ecuadorian obsidian sources used for artifact production and methods for provenience assignments. Latin Am. Antiq **5**, 257 (1994)
2. Dorighel, O., Poupeau, G., Bouchard, J.-F., Labrin, E.: Datation par traces de fission et étude de provenance d'artefacts en obsidienne des sites archéologiques de La Tolita (Equateur. et Inguapi (Colombie). Bull. Soc. Préhist. Fr **91**, 133 (1994)
3. Dorighel, O., Poupeau, G., Labrin, E., Bellot-Gurlet, L.: Fission track dating and provenience of archaeological obsidian artefacts in Colombia and Ecuador. In: Van den Haute, P., de Corte, F. (eds.) Advances in Fission-Track Geochronology, p. 313. Kluwer, Dordrecht (1998)
4. Bellot-Gurlet, L., Poupeau, G., Dorighel, O., Calligaro, T., Dran, J.-C., Salomon, J.: A PIXE/fission-track dating approach to sourcing studies of obsidian artefacts in Colombia and Ecuador. J. Archaeol. Sci **26**, 855 (1999)
5. Scorzelli, R.B., Petrick, S., Rossi, A.M., Poupeau, G., Bigazzi, G.: Obsidian archaeological artefacts provenance studies in the Western Mediterranean basin: an approach by Mössbauer spectroscopy and electron paramagnetic resonance. C. R. Acad. Sci. Paris, Série II **332**, 769 (2001)
6. Duttine, M., Villeneuve, G., Poupeau, G., Rossi, A.M., Scorzelli, R.B.: Electron spin resonance of $Fe^{3+}$ ion in obsidians from Mediterranean islands. Application to provenance studies. J. Non-Cryst. Solids **323**, 193 (2003)
7. Stewart, S.J., Cernicchiaro, G., Scorzelli, R.B., Poupeau, G., Acquafredda, P., De Francesco, A.: Magnetic properties and $^{57}$Fe Mössbauer spectroscopy of Mediterranean prehistoric obsidians for provenance studies. J. Non-Cryst. Solids **323**, 188 (2003)

8. Guillaume-Gentil, N.: Recherches archéologiques sur les monticules artificiels (tolas) dans le nord du bassin du Guayas (Equateur). Modes d'implantation, peuplement, chronologie et échanges interandins. Ph. D. thesis, vol. 2, Université de Neuchâtel (2007)
9. Pereira, C.E., Miekeley, N., Poupeau, G., Küchler, I.L.: Determination of minor and trace elements in obsidian rock samples and archaeological artifacts by laser ablation inductively coupled plasma mass spectrometry using synthetic obsidian standards. Spectrochim. Acta Part B: Atom. Spectrosc **56**, 1927 (2001)
10. Bellot-Gurlet, L., Dorighel, O., Poupeau, G.: Obsidian provenance studies in Colombia and Ecuador: obsidian sources revisited. J. Archaeol. Sci. **35**, 272 (2008)
11. Pereira, C.E.: Otimização de metodologias para análise multielementar: de obsidianas por ICP-MS com amostragem por ablação a laser e aplicações em estudos de proveniência de aterfatos arqueológicos. Ph. D. thesis, Pontifíca Universidade Católica do Rio de Janeiro (2000)

Hyperfine Interact (2007) 175:91–94
DOI 10.1007/s10751-008-9593-0

# Mössbauer investigation of magnetite nanoparticles incorporated in a mesoporous polymeric template

A. F. R. Rodriguez · V. K. Garg · A. C. Oliveira ·
D. Rabelo · P. C. Morais

Published online: 18 March 2008
© Springer Science + Business Media B.V. 2008

**Abstract** Mössbauer spectroscopy was used in this study to investigate magnetite-based nanocomposites, using mesoporous styrene-divinylbenzene (Sty-DVB) microspheres as the hosting template. The magnetite content was increased in the polymeric template by performing several *in situ* chemical reactions (one to six cycles) in the hosting material. We found the Mössbauer linewidth associated to site-A increasing with the relative increase of iron in site-A. The Mössbauer linewidth associated to site-B decreases with the relative decrease of iron in site-B. We explain our findings by assuming local change in the homogeneity associated to changes in the relative iron population in sites A and B.

**Keywords** Magnetite nanoparticle · Polymeric nanocomposite · Mössbauer linewidth

## 1 Introduction

The design and synthesis of nanometer-scaled magnetic structures have recently been the focus of intense investigation, particularly because the unusual or enhanced properties of such materials. As far as the size and size-dispersity control is concerned, tendency of isolated magnetic nanostructures to aggregate into bigger clusters during the synthesis process, driven by particle–particle interaction and/or reduction in energy associated to the high surface-to-volume ratio, has represented a critical obstacle.

Magnetic polymer-based spheres have been considered as an important material in the biotechnology industry, as for instance for cell separation [1] and DNA extraction [2]. In particular, mesoporous polymeric templates could be produced as micron-sized spheres that allow *in-situ* chemical synthesis of nanosized ferrite particles with adjustable magnetic properties and mass density [3]. The fine-tuning of the physical parameters, e.g. net

A. F. R. Rodriguez (✉) · V. K. Garg · A. C. Oliveira · P. C. Morais
Instituto de Física, Núcleo de Física Aplicada, Universidade de Brasília, Brasília 70910-900, Brazil
e-mail: ruiz@unb.br

D. Rabelo
Instituto de Química, Universidade Federal de Goiás, Goiânia 74001-970, Brazil

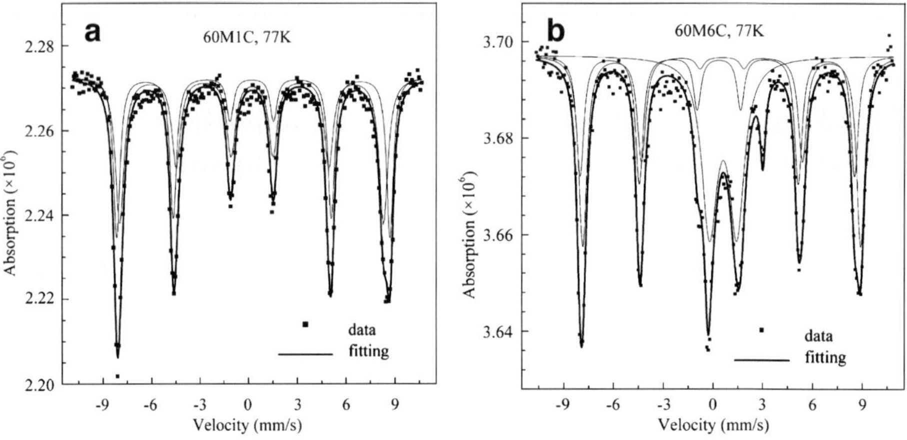

**Fig. 1** Mössbauer spectrum (77 K) of sample **a** 60M1C and **b** 60M6C

magnetic moment and density of the composite can be realized through several cycles ($N$) of chemical synthesis. In the present study Mössbauer spectroscopy was used to characterize the magnetite-based nanoparticle precipitated in mesoporous styrene-divinylbenzene (Sty-DVB) microspheres.

## 2 Experimental

Six magnetite-based composite samples were chemically prepared using the Sty-DVB template immersed in $FeSO_4$ aqueous solution (bath solution) at 60 mmol/L. The Sty-DVB copolymer used was synthesized by suspension polymerization in the presence of inert diluents. The porous structure in the dry polymer template presents an average pore diameter of about 13 nm. The polymeric matrix was characterized through the apparent density (0.44 g/cm$^3$), fixed pore volume (0.44 cm$^3$/g), and surface area (140 m$^2$/g).

The six magnetite-based composite samples were prepared using $N$=1, 2, 3, 4, 5, or 6 chemical cycles of treatment of the polymeric Sty-DVB template. Each chemical cycle ($N$) in the composite synthesis follows a four step procedure. Briefly, the mixture was first stirred for 1 h at room temperature. Second, the iron-loaded polymer particles were separated by filtration and washed thoroughly with water until no ferrous ion was detected in the eluent. Third, the oxidation of the ferrous ion was performed in alkaline medium following a standard recipe to produce magnetite nanoparticles. Fourth, the obtained black composite was filtered, washed with water until the pH of the eluent was neutral, and dried at 60°C. Sample labeling follows uses the general quote 60M$N$C, indicating that the concentration of the bath solution was set at 60 mmol/L (60 mM) whereas $N$ chemical cycles were performed in the synthesis of a particular composite sample. Therefore, the six samples were labeled 60M1C, 60M2C, 60M3C, 60M4C, 60M5C, and 60M6C.

Transmission Mössbauer spectra were recorded at 77 K, using a MCA (256 channels) and a Wissel constant acceleration transducer coupled to a 50 mCi $^{57}$Co/Rh source. Each plastic sample holder (1.7 cm diameter) contained 80 mg of uniformly distributed and pressed sample. The spectra were least square fitted using essentially two sextets. In order to achieve the best curve-fitting procedure doublets have been added to the fitting procedure for three samples (60M3C, 60M4C, and 60M6C). Figure 1(a and b) shows

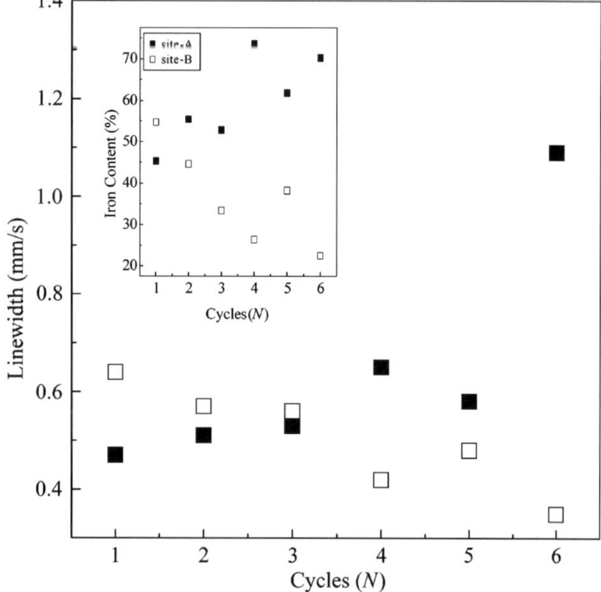

**Fig. 2** Dependence of the Mössbauer linewidth (sites A and B) upon the number of chemical cycles ($N$). The inset shows the dependence of the iron content (sites A and B) upon the number of chemical cycles ($N$) size

typical Mössbauer spectra recorded at 77 K. Figure 2 shows the $N$-dependence (at 77 K) of both the iron content (site-A and site-B) and the Mössbauer linewidth (site-A and site-B).

## 3 Discussion

The values of the internal fields obtained from the fitting of the 77 K Mössbauer spectra were 523 and 508 kOe, assigned to the A (tetrahedral) and B (octahedral) sites, respectively, are in good agreement with the values reported in the literature [4]. At room-temperature the internal field values obtained were 491 kOe (tetrahedral site) and 470 kOe (octahedral). Both set of internal field values (77 and 300 K) indicate the presence of magnetite in the hosting template [4].

The trend shown in the inset of Fig. 2 is quite interesting. As the number of chemical cycles ($N$) increases (from one to six) the relative iron content of site-A monotonically increases with respect to the iron content of site-B. Parallel to the observed iron-site content behavior (see inset of Fig. 2) the Mössbauer linewidth changes systematically as well (see Fig. 2). Data on Fig. 2 show that the Mössbauer linewidth (for both A and B sites) monotonically changes with the relative iron-site occupancy. The physical picture we draw to understand the data presented in Fig. 2 takes into account the magnetite re-growth process that takes place as $N$ increases. Evidences of the re-growth process with changes in the average nanoparticle size and size-dispersity has been previously reported [5]. Also, evidences that the re-growth process may be accomplished through diffusion of chemical species have been reported in the literature [6].

The Mössbauer linewidth provides an indication of the differences and variations of the local chemical environment of iron atoms within the magnetite nanoparticle. The Mössbauer line broadening, or at least its variations with $N$, which includes the influence of the magnetite re-growth process, probably assisted by mass-transfer, may be due to the

presence of slightly different next-nearest neighbor environments in the samples. The systematic change in the relative iron site population as a function of $N$, shifting the sample's composition from close to an inverse cubic spinel structure (for $N=1$), which is the usual bulk magnetite structure, to a site-A iron-enriched structure (for $N=6$) may provide the explanation for our observations. The more the magnetite site-A is iron enriched (relative occupancy), deviating from the usual relative population (2:1), the more site-A local non-homogeneity is created, thus accomplishing the linewidth broadening for site-A, as observed from Fig. 2. In other words, the re-growth process that takes place as a function of $N$ induces iron enrichment (relative content) of site-A, thus resulting in increasing site-A local non-homogeneity. As indicated by the relative site-population, even the very first prepared sample (60M1C) presents an iron over-occupancy of site-A. This indicates that site-B non-homogeneity is already present in the very first composite sample of the series, as indicated by the relative high Mössbauer linewidth value of sample 60M1C (see Fig. 2). Reduction of the Mössbauer linewidth values associated to site-B as $N$ increases indicates that local non-homogeneity is reduced.

## 4 Conclusion

Composite samples containing magnetite nanoparticles immersed in mesoporous styrene-divinylbenzene (Sty-DVB) microspheres were investigated by Mössbauer spectroscopy. Samples with different magnetite contents were produced making use of cycles of chemical incorporation of the magnetic material in hosting template. The Mössbauer spectra recorded from a series of these samples provide evidences of variations of the relative iron-site occupancy as a function of the number of chemical cycles. For site-A we found the Mössbauer linewidth increasing with the increase of the relative iron content. We interpret this finding assuming that site-A non-homogeneities increases as the relative iron content increases. In contrast, the relative reduction of iron content in site-B reduces the Mössbauer linewidth, indicating the enhancement of the local homogeneity.

**Acknowledgment** This work was supported by the Brazilian agency CNPq/MCT and FINATEC.

## References

1. Chatterjee, J., Haik, Y., Chen, C.-J.: J. Magn. Magn. Mater. **225**, 21 (2001)
2. Oster, J., Parker, J., Brassard, L.: J. Magn. Magn. Mater. **225**, 145 (2001)
3. Rabelo, D., Lima, E.C.D., Reis, A.C., Nunes, W.C., Novak, M.A., Morais, P.C.: Nano Lett. **1**, 105 (2001)
4. Kuzmann, E., Nagy, S., Vértes, A., Weiszburg, T.G., Garg, V.K.: In: Vertes, A., Nagy, S., Süvegh, K. (eds.) Geological and mineralogical applications of Mössbauer spectroscopy in nuclear methods in mineralogy and geology: techniques and applications, pp. 285–376. Plenum Press, New York (1998)
5. Morais, P.C., Azevedo, R.B., Silva, L.P., Rabelo, D., Lima, E.C.D.: Phys. Status Solidi (A) **187**, 203 (2001)
6. Morais, P.C., Azevedo, R.B., Silva, L.P., Lacava, Z.G.M., Lima Filho, F.P., Lemos, A.P.C., Rabelo, D., Barbosa, D.P., Lima, E.C.D.: Phys. Status Solidi (A) **201**, 898 (2004)

Hyperfine Interact (2007) 175:95–101
DOI 10.1007/s10751-008-9594-z

# Distribution of Fe-bearing compounds in an Ultisol as determined with selective chemical dissolution and Mössbauer spectroscopy

C. Pizarro · J. D. Fabris · J. Stucki · V. K. Garg ·
C. Morales · S. Aravena · J. L. Gautier · G. Galindo

Published online: 5 June 2008


**Abstract** The distribution of Fe-bearing compounds in the silt and clay fractions from samples of a Chilean Ultisol developing on pyroclastic material was determined in this study. The mineralogical analysis was carried out mainly by Mössbauer spectroscopy at 298, 80 and 6 K, for samples before and after treatment with ammonium oxalate (OX). The 298 and 80 K-Mössbauer spectra reveal relatively complex mineral assemblages for all samples. An intense central doublet, due to (super)paramagnetic $Fe^{3+}$ is observed. The incipient broad-line sextet may be due to coexisting magnetically ordered forms of iron oxides. Goethite, hematite and maghemite were identified from the Mössbauer spectral analysis of patterns obtained at 6 K, which were numerically fitted with model-independent hyperfine field distributions. Some low hyperfine fields are identified for both, the silt and clay samples, being more intense in the last one. These species are likely due to poorly crystalline iron oxyhydroxides, which are more easily removed with the OX treatment; maghemite tends to remain in the coarse fraction, whereas poorly crystalline forms do occur mainly in the clay fraction.

**Keywords** Volcanic soils · Iron oxides · Ammonium oxalate · Mössbauer spectroscopy

C. Pizarro (✉) · C. Morales · S. Aravena · J. L. Gautier · G. Galindo
Facultad de Química y Biología, Universidad de Santiago de Chile, Av. B. O'Higgins. 3363,
Casilla 40 Correo 33, Santiago, Chile
e-mail: cpizarro@lauca.usach.cl

J. D. Fabris
Departamento de Química, ICEx, Universidade Federal de Minas Gerais, Pampulha,
31270–901 Belo Horizonte, Minas Gerais, Brazil

J. Stucki
NRES/ACES, University of Illinois, Urbana, IL 61801, USA

V. K. Garg
Instituto de Física, Universidade de Brasília, 70910-900 Brasília, Distrito Federal, Brazil

# 1 Introduction

The iron oxides (a broad term, meaning here iron hydroxide, oxyhydroxide or oxide) particles, part of them having a variable structural electric charge, are major components in many acidic soils and greatly influence the ion dynamic in the solid-solution relationship. In general, the ion sorption rate in soils depends on the surface area, chemical composition, cation exchange capacity, and on the proportion of iron or aluminum oxides or oxyhydroxides present as the surface coating of clay and silt particles. In this sense, the way these minerals are inter-related or chemically spatially distributed, either being freely distributed throughout the soil mass or coating silt and clay grains, is determinant on their chemical role in the whole ion sorption-desorption mechanisms.

Soils derived from volcanic materials are abundant and widespread in southern Chile. They are mostly found from latitude 19° to 56° S, and account for around 70% of the Chilean cultivated areas. The Ultisols (classification according to the U.S. Soil Taxonomy [1, 2]) are one of the most important volcanic Chilean soils. They have relatively high amounts of iron oxides in different degrees of crystallinity, low amounts of organic matter, and mineralogy mainly dominated by kaolinite and halloysite.

Physical extraction methods and selective chemical dissolution are useful laboratory procedures in the sample preparation that improve the mineralogical analysis of soil materials [3]. The oxalic acid—ammonium oxalate treatment is conventionally used to extract poorly crystalline Fe oxides. In this work, the characterization and distribution of Fe-compounds in the silt (mean diameter of particles, $\varphi=0.002$–$0.053$ mm) and clay ($\varphi<0.002$ mm) fractions from soil volcanic Chilean samples were studied.

# 2 Materials and methods

## 2.1 Soil collection and fraction separations

Soil samples collected at a depth of 15–30 cm from an Ultisol profile (geographical coordinates of the sampling site 36°58′ S 72°09′ W) in Collipulli, southern Chile, were selected. The collected soil material was sieved in the field, so to obtain a fraction with mean particle diameter of 2 mm, and packed under roughly the corresponding field moisture atmosphere. The particle size fractionation was first carried out by dispersing an aliquot of the soil sample in distillate water. The soil samples were first dispersed in distillate water; the suspension was then sieved, to separate the sand fraction ($\varphi=2$–$0.05$ mm). The silt ($\varphi=0.05$–$0.002$ mm) and the clay fraction ($\varphi<0.002$ mm) were separated by sedimentation according the method described in [4].

## 2.2 Mineralogical analysis

The powder X-ray patterns were obtained with a Philips Norelco Instrument equipped with a Cu-K$\alpha$ radiation source and Ni filter. The Mössbauer spectra were collected with the sample at room temperature (298 K), 80 and 6 K in a constant acceleration transmission set up with a ×30 mCi $^{57}$Co/Rh source. The specific soil magnetization was measured with a portable soil magnetometer developed and built by Coey et al. [5], which allows direct digital readings of the magnetic moment of soil samples expressed in microjoule per tesla; the specific saturation magnetization (*SM*) in joule per tesla per kilogram was deduced from

 Springer

**Table 1** Chemical composition, OX-extraction and organic carbon (O.C.) of the soil, silt and clay Collipulli samples

| Sample | $Al_2O_3$ (wt.%) | $Fe_2O_3$ (wt.%) | $SiO_2$ (wt.%) | $TiO_2$ (wt.%) | $Fe_{OX}$ (wt.%) | O.C. (wt.%) |
|---|---|---|---|---|---|---|
| Soil | 23.2 | 15.6 | 40.6 | 2.3 | 2.0 | 2.4 |
| Silt | 23.1 | 14.0 | 46.6 | 2.3 | 1.2 | 1.2 |
| Clay | 30.2 | 11.6 | 37.7 | 1.2 | 0.8 | 1.2 |

**Table 2** Granulometric fractioning and values of specific saturation magnetization (*SM*)

| Sample | Proportion/ wt.% | Specific saturation magnetization, $SM$/J T$^{-1}$ kg$^{-1}$ |
|---|---|---|
| Soil |  | 2.4 |
| Sand | 10.4 | 1.7 |
| Silt | 37.6 | 1.4 |
| Clay | 52.0 | 1.2 |

the sample mass. About 50 mg were used to obtain an averaged *SM* value from 20 readings per sample.

## 2.3 Chemical analyses and dissolutive methods

Total concentrations were determined by digesting 0.100 g of the dry sample with HF and aqua regia (10:1.5) in a Parr bomb at 110°C during 12 h. The resultant digestion mixture was treated with ultra pure grade (Merck) boric acid and diluted to 100 mL with water, and stored in a plastic bottle at 5°C, prior to the chemical analysis. Data in Table 1 are averaged values for two sample replicas.

Each clay and silt fractions of the Ultisol volcanic soil samples were chemically treated with oxalic acid-ammonium oxalate (OX) during 4 h at pH 3, in darkness, to extract the poorly crystallized iron oxides, according to procedures described in [6]. After this extraction and total dissolution procedures, the limpid supernatant was separated, by centrifuging at 10,000 rpm, and analyzed for Fe, Al and Si, with ICP-OES. The organic matter content was determined by the Walkley-Black method [7], using a Radiometer automatic titrator provided with a Pt-calomel combined electrode. The chemical composition of untreated and OX-treated samples and organic carbon contents are presented in Table 1. The pH of samples was measured in soil suspension prepared with double distilled water at a ratio of 1:2.5 *w/v*. The granulometric distribution was determined according to the pipette method, based in the sedimentation velocity Stoke's law.

## 3 Results and discussion

The chemical composition of samples from the whole soil and its silt and clay fractions indicates that this Ultisol does contain a high proportion of iron oxyhydroxides, and is

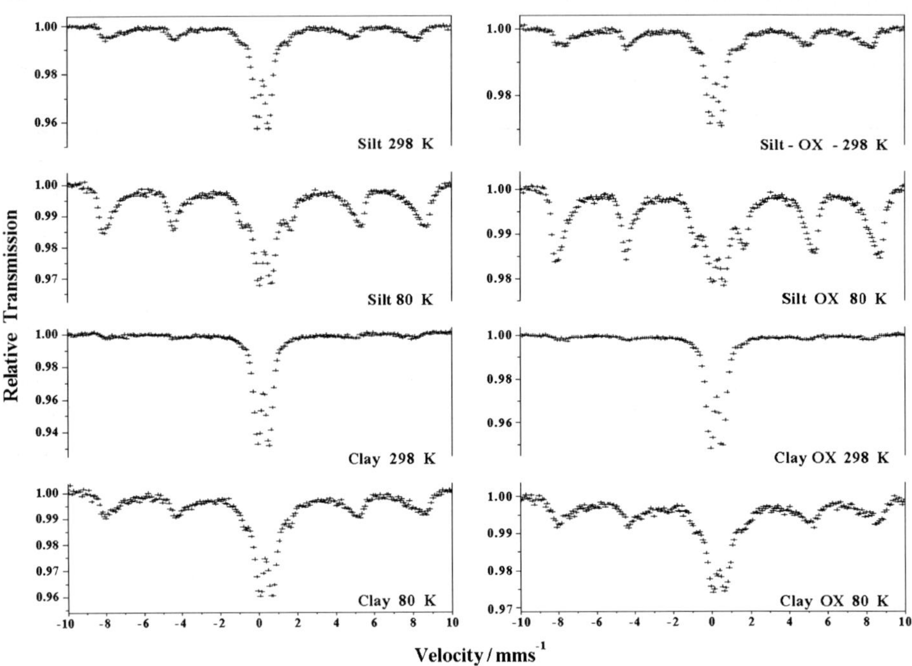

**Fig. 1** 298 and 80 K Mössbauer spectra for Collipulli silt and clay fractions, before and after of the OX treatment

relatively poor in organic matter (Table 1). This soil presents high clay content (Table 2). All these characteristics are similar to those found for other volcanic Chilean Ultisols [8]. This clay fraction was found to contain mainly kaolinite (>50 wt.%) along with $\alpha$-cristobalite, halloysite, vermiculite at <1 wt.% [9]. The magnetic measurements (Table 2) indicated this is a magnetic soil.

For both, untreated and OX-treated silt and clay fractions, the 298 and 80 K-Mössbauer (Fig. 1) spectra reveal relatively complex mineral assemblages, with heterogeneous chemical compositions, rendering patterns mainly formed by an intense central doublet, due to paramagnetic $Fe^{3+}$ in silicates and/or superparamagnetic iron oxides. The incipient broad-line sextet may be due to coexisting magnetically ordered forms of iron oxide. The sextet signals are more intense for the silt than for the clay sample. Goethite, hematite and maghemite were identified from the Mössbauer spectral analysis of patterns obtained at 6 K, for both particle size fractions, (Fig. 2). The relative area for maghemite in non-treated samples was found to be 46%, for the silt, and 13%, for the clay. These results are consistent with the measured values of specific saturation magnetization for silt and clay fractions (Table 2). Some low hyperfine fields are identified for both samples: (1) maximum probability at 44.7 T, accounting for about 16% of the relative spectral area ($RA$) of the silt, but more intense for the clay sample; (2) 44.2 T and 47.3 T, summing 31% $RA$ (Table 3). These magnetically ordered species are likely due to poorly crystalline iron oxyhydroxides, which are more easily removed with the OX treatment (Fig. 2). In this volcanic soil, maghemite tends to remain in the coarse fraction, whereas poorly crystalline structures do occur mainly in the clay fraction (Table 4).

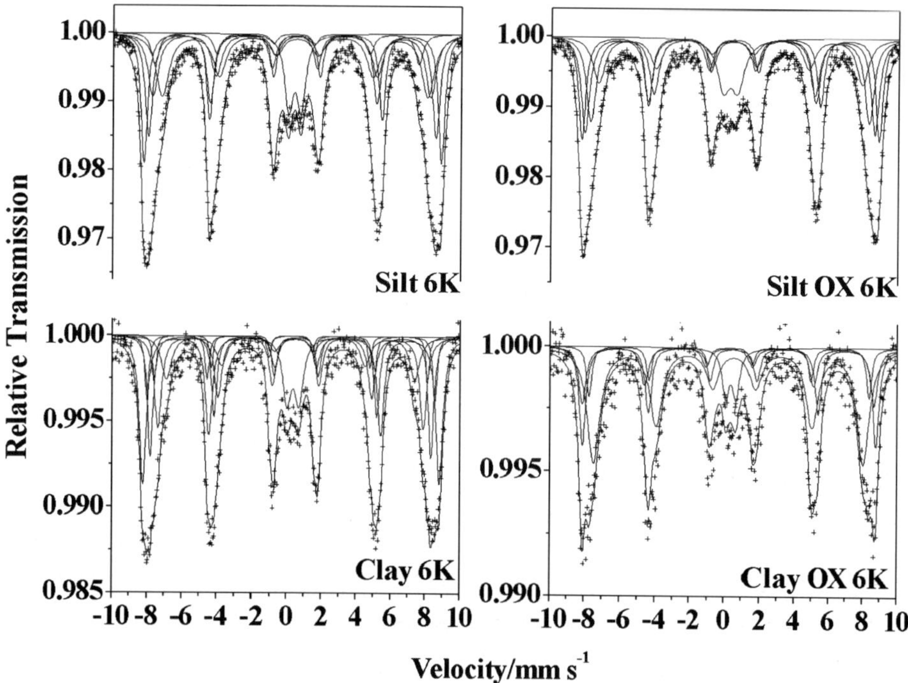

**Fig. 2** 6 K Mössbauer spectra for Collipulli silt and clay fractions, before and after of the OX treatment

**Table 3** 6 K Mössbauer parameters for the UT Collipulli and for the ammonium OX silt samples

| Sample | Subspectrum | $\delta$/mm s$^{-1}$ | $\Delta,\varepsilon$/mm s$^{-1}$ | $B_{hf}$/T | $RA$/% |
|---|---|---|---|---|---|
| 6K-UT | Fe(III) | 0.32(1) | 0.91(2) | | 11.8(2) |
| | Gt | 0.57(1) | −0.35(1) | 50.1(1) | 11(3) |
| | Hm | 0.54(1) | −0.21(1) | 53.3(1) | 16(3) |
| | [Mh] | 0.45(1) | −0.08(2) | 48.3(1) | 21(5) |
| | {Mh} | 0.43(1) | −0.04(1) | 51.8(1) | 25(3) |
| | Lower Field | 0.53(2) | −0.19(4) | 44.7(4) | 16(3) |
| 6K-OX | Fe(III) | 0.39(1) | 0.83(4) | | 13.7(2) |
| | Gt | 0.50(1) | −0.20(3) | 49.6(2) | 24(8) |
| | Hm | 0.52(1) | −0.17(1) | 53.0(1) | 22(6) |
| | [Mh] | 0.45(2) | −0.07($^{a}$) | 47.0(5) | 17(5) |
| | {Mh} | 0.44(1) | 0.06(2) | 51.7(2) | 24(8) |

[] and {} stand, respectively, for iron in tetrahedral and octahedral sites of maghemite. Numbers in parentheses refer to errors over the last significant digits as given by the least-squares procedure and may not represent the actual experimental uncertainties

*UT* Untreated, *OX* oxalate-treated $\delta$ isomer shift relative to $\alpha$Fe; $\varepsilon$ quadrupole shift, $B_{hf}$ hyperfine field, *RA* relative area of the subspectrum, *Fe(III)* subspectrum due to (super)paramagnetic Fe$^{3+}$, *Gt* goethite, *Hm* hematite, *Mh* maghemite, *Lower Field* any not-assigned subspectrum, for which the hyperfine field value is lower than to those of the previous magnetically ordered phases

$^{a}$ Fixed value during least-squares fitting convergence

**Table 4** Mössbauer parameters for the UT Collipulli and for the ammonium OX clay samples

| Sample | Subspectrum | $\delta$/mm s$^{-1}$ | $\Delta,\varepsilon$/mm s$^{-1}$ | $B_{hf}$/T | $RA$/% |
|---|---|---|---|---|---|
| 6K-UT | Fe(III) | 0.45(2) | 0.70(2) | | 11.2(4) |
| | Gt | 0.51(1) | −0.26(2) | 50.0(1) | 18(4) |
| | Hm | 0.50(1) | −0.20(1) | 52.8(1) | 28(3) |
| | [Mh] | 0.37(1) | −0.03(3) | 48.7(3) | 5(1) |
| | {Mh} | 0.38(1) | −0.03(2) | 51.3(1) | 8(1) |
| | Lower Field | 0.44(2) | −0.16(4) | 44.2(3) | 12(3) |
| | Lower Field | 0.49(1) | −0.23(2) | 47.3(1) | 19(4) |
| 6K-OX | Fe(III) | 0.50($^a$) | 0.60($^a$) | | 10.3(6) |
| | Gt | 0.52(2) | −0.28($^a$) | 48.1(2) | 30(4) |
| | Hm | 0.53(2) | −0.18($^a$) | 52.3(1) | 32(5) |
| | [Mh] | 0.43(3) | 0.0($^a$) | 49(1) | 11(2) |
| | {Mh} | 0.38(1) | 0.0($^a$) | 52.0(2) | 18(2) |

[] and {} stand, respectively, for iron in tetrahedral and octahedral sites of maghemite. Numbers in parentheses refer to errors over the last significant digits as given by the least-squares procedure and may not represent the actual experimental uncertainties

*UT* Untreated, *OX* oxalate-treated, $\delta$ isomer shift relative to $\alpha$Fe, $\varepsilon$ quadrupole shift, $B_{hf}$ hyperfine field, *RA* relative area of the subspectrum, *Fe(III)* subspectrum due to (super)paramagnetic $Fe^{3+}$, *Gt* goethite, *Hm* hematite, *Mh* maghemite, *Lower Field* any not-assigned subspectrum, for which the hyperfine field value is lower than to those of the previous magnetically ordered phases

$^a$ Fixed value during least-squares fitting convergence

## 4 Conclusions

Goethite, hematite and maghemite were identified in the silt and clay fractions of this Collipulli volcanic Ultisol. Maghemite tends to remain in the coarse fraction, whereas poorly crystalline forms do occur mainly in the clay fraction. The selective chemical treatment revealed to be an essential step of sample preparation for the mineralogical analysis with physical methods.

**Acknowledgments** This work was supported by http://www.fondecyt.cl/Fondo Nacional de Desarrollo Científico y Tecnológico (Chile) 1040272 and 1050178, MECESUP USA 9903 (Chile), Conselho Nacional de Desenvolvimento Científico e Tecnológico (Brazil), Fulbright (USA)/Coordenação de Aperfeiçoamento de Pessoal de Nível Superior (Brazil), and the Electronic Retailing Self-Regulation Program, U.S. Department of Energy/Bureau for Economic Reseiarch, Grant N° DE-FG02-00ER62986 (subcontract FSU F48792); and NSF Grant N° EAR 01-26308.

## References

1. U.S. Department of Agriculture, Natural Resources Conservation Service: National Soil Survey Handbook, title 430-VI. Available at: http://soils.usda.gov/technical/handbook/, (2005)
2. U.S. Department of Agriculture, Natural Resources Conservation Service: The twelve orders of Soil Taxonomy. Available at: http://soils.usda.gov/technical/soil_orders/, (2006)
3. Lelis, M.F.F., Gonçalves, C.M., Fabris, J.D., Mussel, W.N., Pacheco Serrano, W.A., Braz, J.: Chem. Sci. **15**, (6), 884–889 (2004)
4. Jackson, M.L.: Soil Chemical Analysis—Advanced Course. Madison Libraries, Madison (2)(1975)

5. Coey, J.M.D., Cugat, O., McCauley, J., Fabris, J.D.: A portable soil magnetometer. Rev. Apl. Instrumfs. **7**, (No 1), 25–29 (1992)
6. Schwertmann, U.: Z. Pflanzenernähr. Düng. Bodenkd. **84**, 194 (1959)
7. Allison, L.E.: In: Black, C.A., Evans, D.D., White, J.L., Ensminger, L.E., Clark, F.E. (eds.) Agronomy 9, Part 2, p. 1367. American Society of Agronomy, Madison, W.I. (1965)
8. Escudey, M., Galindo, G., Förster, J.E., Briceño, M., Díaz, P., Chang, A.: Commun. Soil Sci. Plant Anal. **32**, (5,6), 606 (2001)
9. Pizarro, C., Escudey, M., Fabris, J.D.: Hyp. Int. **148/149**, 53–59 (2003)

Hyperfine Interact (2007) 175:103–111
DOI 10.1007/s10751-008-9595-y

# Structural, magnetic and hyperfine characterization of zinc-substituted magnetites

A. C. S. da Costa · I. G. de Souza Jr. · M. A. Batista ·
K. L. da Silva · J. V. Bellini · A. Paesano Jr.

Published online: 6 May 2008
© Springer Science + Business Media B.V. 2008

**Abstract** Nanosized $Fe_{3-x}Zn_xO_4$ powders were synthesized by co-precipitation and characterized by total chemical analysis, X-ray diffraction, magnetic susceptibility and Mössbauer spectroscopy. The results showed that, for $x \leq 0.15$, the as-prepared samples are mostly zinc-substituted magnetites but have maghemite as a minor phase. For $x \geq 0.30$, only the $Fe_{3-x}Zn_xO_4$ solid solution is found. Increasing the zinc content from the end concentration $x=0$, increases the lattice parameter but smaller become the mean crystalline diameter and the magnetic susceptibility. In addition, the magnetic hyperfine fields of the iron sites in the spinel structure, A and B, decrease up to collapse at $x \leq 0.90$.

**Keywords** Spinels · Isomorphous substitution · Zn-substituted · Mössbauer spectroscopy

## 1 Introduction

Ferrimagnetic oxides are intensely investigated because of their important applications to the industry [1] and the environment [2]. In soils and sediments, magnetite ($Fe_3O_4$) and maghemite ($\gamma$-$Fe_2O_3$) are the most common species belonging to that class of oxides [3].

Naturally occurring magnetites and maghemites are frequently intergrown with other minerals or form highly stable aggregates, respectively, making their separation very difficult [3]. In rocks, soils and sediments the ferrimagnetic minerals do not present solely Fe, O, and OH, but isomorphous substitution is a rule among theses minerals. The most common metals substituting for iron are $Ti^{4+}$, $Al^{3+}$, $Zn^{2+}$, $Cu^{2+}$, $Ni^{2+}$, $Mn^{2+}$ and $Co^{2+}$ [4]. The presence of these metals not only affects the chemistry and mineralogical attributes but also the magnetic behavior of those minerals [3, 4].

A. C. S. da Costa (✉) · I. G. de Souza Jr. · M. A. Batista
Departamento de Agronomia, Universidade Estadual de Maringá, 87020-900 Maringá, PR, Brazil
e-mail: acscosta@uem.br

K. L. da Silva · J. V. Bellini · A. Paesano Jr.
Departamento de Física, Universidade Estadual de Maringá, Maringá, PR, Brazil

Substituted magnetites have also been synthesized by the sol-gel combustion method [5, 6], ball milling procedure [7, 8], and electrochemical processes [9]. Recently, special attention has been focused on the synthesis methods based on co-precipitation from aqueous solutions [10–14], with different cations substituting the iron in the spinel structure, and on how this affects its magnetic properties.

The iron for zinc substitution in magnetites originates a solid solution with end members magnetite ($Fe_3O_4$) and franklinite ($ZnFe_2O_4$), both with spinel structure. Magnetite has an inverse spinel structure where the trivalent Fe(III) fully occupies the A-site and the remaining Fe(III) and Fe(II) occupy the B-sites. Franklinite is an anti-ferromagnetic oxide, in which zinc occupies de A tetrahedral site and Fe(III) the octahedral B-sites. Franklinite may also have a mineral origin and it is usually found in the exploration of zinc mines.

In this work we characterize zinc-substituted magnetites synthesized by co-precipitation. The aim is to determine the structural, magnetic and hyperfine properties of the $Fe_{3-x}Zn_xO_4$ solid solutions, as the zinc concentration is varied.

## 2 Experimental details

### 2.1 Synthesis method

Magnetites were synthesized following, with modifications, the Schwertmann and Cornell [15] procedure. $FeSO_4$ and $ZnSO_4$ solutions were dissolved in water in different iron to zinc proportions, to nominally attain the x proportions (0, 0.075, 0.15, 0.30, 0.45, 0.60, 0.90) in the general chemical formula $Fe_{3-x}Zn_xO_4$. The reaction was kept under hot (90°C) and $N_2$ atmospheric conditions. With the addition of an alkali (KOH 1M) to the solution, a bluish material easily precipitated that evolved to a dark powder within less than 1 hour and was easily attracted to a hand magnet. Exception was the treatment for the x=0.9 sample, that was kept warm and under $N_2$ atmosphere for almost one day. All synthetic materials were washed free of salts with de-ionized water for several days, frozen under liquid nitrogen and freeze-dried.

### 2.2 Characterization

The chemical compositions of the powder materials were verified after dissolution with hot (75°C) concentrated (1:1 *v/v*) $H_2SO_4$. Iron and zinc, in solution, were determined with a GBC atomic absorption spectrometer. The measured concentrations confirmed within 5% of error the initial. The XRD analysis was carried on a Shimadzu D6000 equipment, having a Cu-K$\alpha$ ($\lambda$=0.15418 nm) beam and a Ni filter. The XRD scanning step was 0.02° (2$\theta$), with 0.6 s for the time of pulse accumulation. The X-ray diffraction patterns were used to evaluate the unit cell parameter, *a*, and the mean crystal diameter size (MCD), *d*, based on Scherrer's formula. Mass specific ($\chi_{lf}$) and frequency dependent ($\chi_{fd}$) magnetic susceptibility were measured for each powder material using a Bartington MS2 Magnetic Susceptibility System (Bartington Instruments LTD) coupled with a MS2B sensor, equipped with both low frequency (0.46 kHz) and high frequency settings (4.65 kHz) for the identification of fine grained superparamagnetic materials. The Mössbauer spectroscopy (MS) characterizations were performed in the transmission geometry, using a conventional Mössbauer spectrometer, in a constant acceleration mode. The $\gamma$-rays were provided by a $^{57}$Co(Rh) source. The Mössbauer spectra were analyzed with a non-linear least-square routine, with Lorentzian line shapes. Hyperfine magnetic field distributions were

 Springer

occasionally traced by means of histograms, with a fixed linewidth ($\Gamma$). All isomer shift (IS) data given are relative to $\alpha$-Fe throughout this paper.

## 3 Results and discussion

Figure 1 presents the X-ray diffraction patterns for some selected samples. The diffractograms show planes that can be indexed to a cubic spinel structure (PDF# 19-0629), revealing the formation of magnetites or, eventually, maghemites. However, differentiation between zinc-substituted magnetite and maghemite by ordinary X-ray diffraction is difficult, if not impossible, due to the X-ray scattering ability of the metal cation (Zn), since it is close to the iron element [8].

A meticulous analysis of peaks positions reveals that those lines of the planes parallel to the direction <100> [i.e., the (h,k,0) planes] shift to lower angles, increasing the interplane

**Fig. 1** X-ray diffraction patterns for the Zn-substituted magnetites

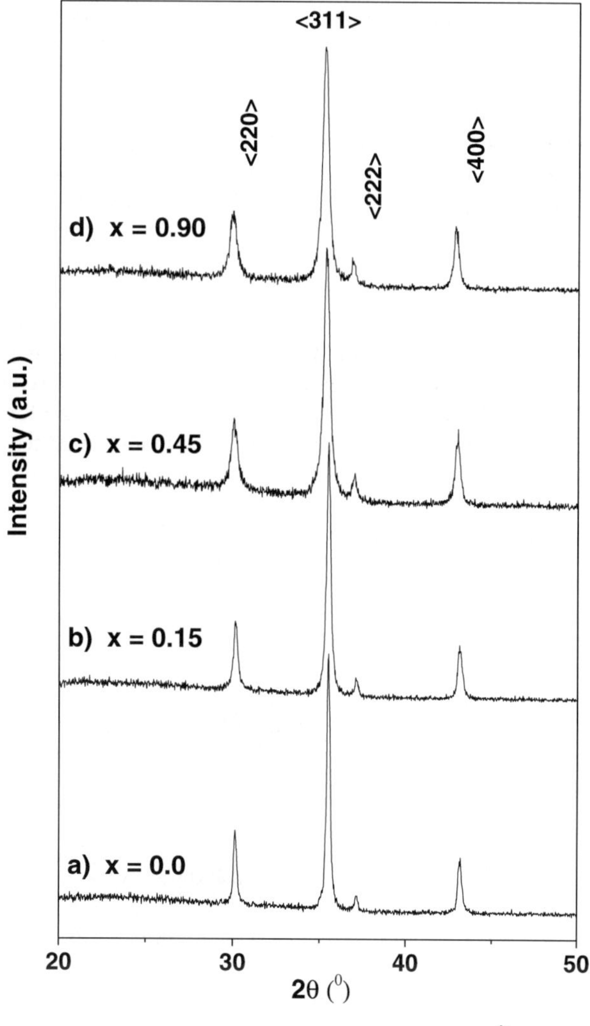

distance and, consequently, the lattice parameter, *a,* as the degree of iron for zinc substitution increases [8]. Other planes (e.g., the (311), (511) and (533) planes, here designated according to the spinel cubic structure) show first to displace to right, then to left, which may indicate a fluctuation in the interplane distance. This could be due to a tetrahedral distortion from the cubic symmetry as a consequence of the zinc substitution. New measurements, statistically more significant in terms of X-ray profile, are currently being conducted together with a Rietveld analysis and shall be reported elsewhere. In any case, at the end concentrations all planes yield close values for the lattice parameter.

Figure 2 shows the variation of the mean lattice parameter, averaged over the (hk0) planes, as a function of the zinc content. The mean lattice parameter varies from 8.362 Å ($x$=0.0) up to 8.447 Å ($x$=0.90), which are consistent values considering that $a$ is 8.396 Å for magnetite (PDF # 19-02629) and is 8.441 Å for franklinite (PDF # 22-1012). As noted above, these numbers are also matched by the planes l$\neq$0.

The variation in the mean lattice parameter can be understood as follow: the ionic radius of $Zn^{2+}$ in tetrahedral coordination (0.74 nm) is similar to $Fe^{2+}$ in octahedral coordination (0.75 nm); however, the "transfer" of $Fe^{3+}$ from tetrahedral coordination (0.63 nm) to octahedral coordination (0.78 nm) produces a net increase in the unit cell parameter. The opposite trend is observed when $Al^{3+}$ is the substituting cation, due to its smaller ionic radii (0.54 nm) [1, 12].

A detailed inspection of the diffractograms in the Fig. 1 reveals, in addition, changes in the width of the diffraction lines with increasing values of x. The mean crystalline diameter, *d,* calculated from the full width at half-height of each diffraction line, decreases monotonically with increasing iron for zinc substitution [3, 9, 11, 14], as shown in Fig. 3.

Actually, crystallites formed with isomorphous substitution from co-precipitation methods usually have smaller particle size associated with the ill development of some planes and or crystallographic axis [3, 15, 17], as it is the case for the present magnetites.

The mass specific magnetic susceptibility ($\chi_{lf}$), as a function of $x$, is shown in Fig. 4. It can be seen that increasing the content of the $Zn^{2+}$ diamagnetic cation reduces the overall magnetization of the magnetite [3]. This was already observed previously for $Ti^{4+}$, $Al^{3+}$ and $Mg^{2+}$ [13, 15–17]. The magnetic susceptibility obtained for the $Fe_3O_4$ (i.e., 55,000 $10^{-8}$ $m^3$ $Kg^{-1}$) is in reasonable agreement with previously reported values [3, 16, 18]. In the other

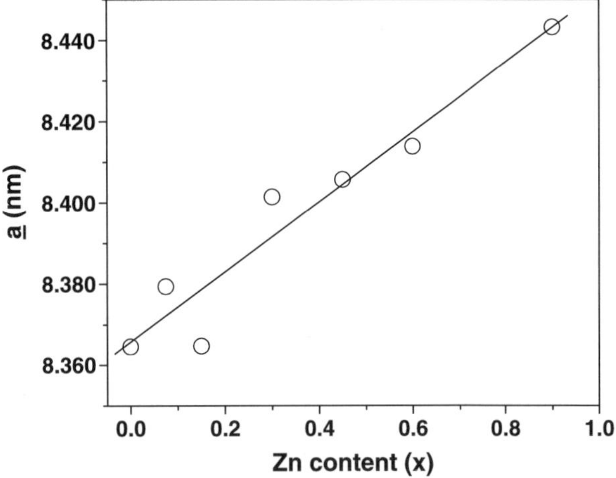

**Fig. 2** Average lattice parameter, *a*, for the Zn-substituted magnetites, as a function of *x*

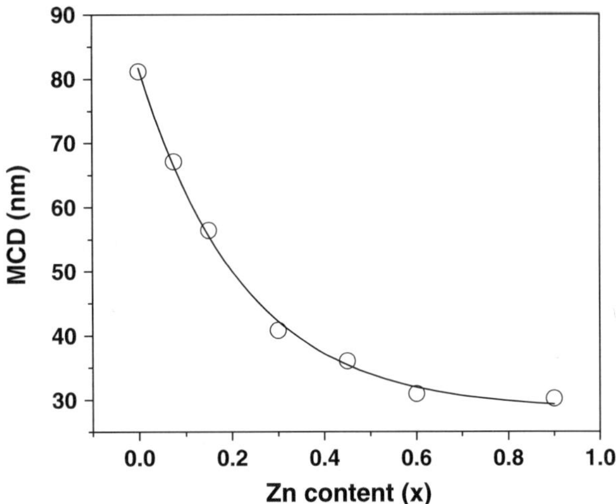

**Fig. 3** Mean crystalline diameter for the Zn-substituted magnetites, as a function of *x*. The continuous line is only a guide to eye

extreme, the zinc richest sample, although with composition close to franklinite, still shows a ferrimagnetic character, according to its $\chi_{lf}$ value ($\sim$30,000 $10^{-8}$ m$^3$ Kg$^{-1}$).

The frequency dependence of the mass specific magnetic susceptibility ($\chi_{fd}$, %) increases with *x* values (data not shown) and is inversely related to the mean crystalline diameter, *d* [16].

Soils and sediments show this particular behavior and smaller values of $\chi_{fd}$ (<10) [16, 18] are observed in the sand and silt fractions, 2 to 0.002 mm diameter, due to the presence of coarse single domain and multidomain grains of ferrimagnetic minerals. In the clay size fraction, diameter<2 µm, small superparamagnetic crystallites are majority, increasing $\chi_{fd}$ values [16, 17].

Figure 5 shows the Mössbauer spectra for some selected samples. The hyperfine parameters and subspectral areas are given in Table 1.

For the lower concentrations (i.e., for $x \leq 0.15$), the best fits are obtained using two discrete sextets and a hyperfine magnetic field distribution (Fig. 5a). Based on the X-ray

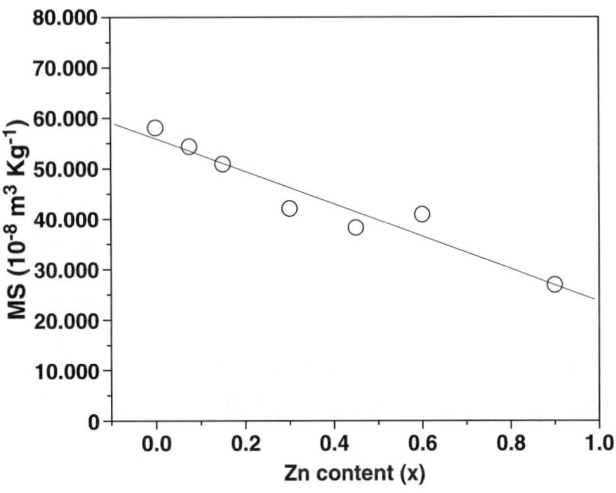

**Fig. 4** Mass specific magnetic susceptibility for the Zn-substituted magnetites, as a function of *x*

 

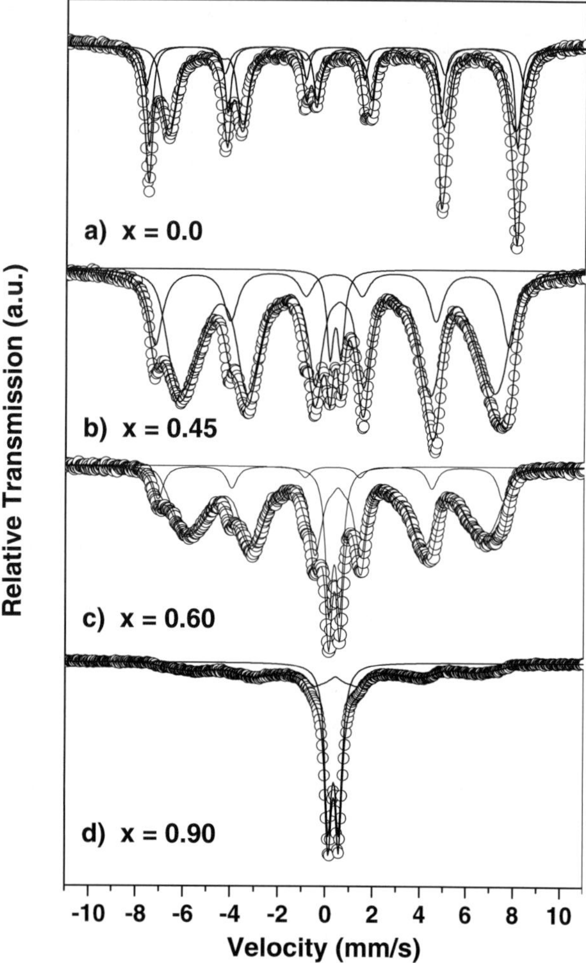

**Fig. 5** RT Mössbauer spectra for the $x$=0.0 (**a**), 0.45 (**b**), 0.60 (**c**) and 0.90 (**d**) for the Zn-substituted magnetites



results and previously established Mössbauer parameters, the smallest sextet and the distribution may be attributed, respectively, to the sites A (tetrahedral) and B (octahedral) of the $Zn_xFe_{3-x}O_4$ magnetite phase [3, 4, 10, 15]. Still for the lower range of $x$ values, the area ratios are close to 1:2, as expected for nearly stoichiometric magnetites.

The line broadening showed by the site B is related to the weakening of the hyperfine field, due to small sized crystallites, and/or non-magnetic substitutional cations (e.g., Zn cations) in the magnetite structure [10, 13]. It is well known that $Zn^{2+}$ ions have preference for the A-site [8, 14], while other transition metals (Co, Ni, Mn, Cu and Cr) prefer the B-site [11, 12].

On the other hand, the largest sextet has hyperfine parameters comparable with those reported for maghemite, also a spinel-like mineral [3, 4, 10, 15]. For this reason, a clear separation between the diffraction peaks of both phases is not expected to be visible in ordinary X-ray patterns for samples having, mixed, magnetite and maghemite (Fig. 1), as pointed out above.

Maghemite was probably formed by oxidation of the very fine magnetite particles occurring in the as-prepared samples [10], since no special storing procedure was destined

**Table 1** Mössbauer hyperfine parameters and subspectral areas for the $Fe_{3-x}Zn_xO_4$ magnetites

| Sample (x) | Phase/site | | $IS^a$ (mm/s) (±0.02) | QS (mm/s) (±0.02) | $B_{hf}^b$ (T) (±0.2) | $\Gamma^c$ (mm/s) (±0.2) | Area (%) (±0.3) |
|---|---|---|---|---|---|---|---|
| 0.00 | Magnetite | Site A | 0.28 | −0.03 | 48.3 | 0.31 | 27.7 |
| | | Site B | 0.65 | −0.03 | 43.4 | – | 54.7 |
| | Maghemite | | 0.32 | 0.04 | 49.9 | 0.47 | 17.6 |
| 0.075 | Magnetite | Site A | 0.31 | −0.04 | 48.4 | 0.37 | 28.9 |
| | | Site B | 0.53 | −0.12 | 41.4 | – | 58.0 |
| | Maghemite | | 0.29 | 0.05 | 49.8 | 0.31 | 13.1 |
| 0.15 | Magnetite | Site A | 0.30 | −0.03 | 48.0 | 0.41 | 28.9 |
| | | Site B | 0.56 | −0.09 | 39.8 | – | 59.3 |
| | Maghemite | | 0.30 | 0.04 | 49.4 | 0.34 | 11.8 |
| 0.30 | Magnetite | Site A | 0.28 | −0.03 | 47.5 | 0.41 | 26.6 |
| | | Site B | 0.54 | −0.05 | 39.9 | – | 69.7 |
| | | | 0.36 | 0.47 | – | 0.31 | 3.7 |
| 0.45 | Magnetite | Site A | 0.28 | −0.03 | 46.5 | 0.72 | 22.4 |
| | | Site B | 0.52 | −0.03 | 36.8 | – | 71.9 |
| | | | 0.36 | 0.46 | – | 0.34 | 5.7 |
| 0.60 | Magnetite | Site A | 0.29 | −0.02 | 45.2 | 0.63 | 15.2 |
| | | Site B | 0.49 | −0.03 | 33.1 | – | 70.1 |
| | | | 0.35 | 0.46 | – | 0.35 | 14.7 |
| 0.90 | Magnetite | Sites A,B | 0.41 | −0.02 | 26.8 | – | 52.3 |
| | | Site B | 0.35 | 0.44 | – | 0.35 | 47.7 |

$^a$ Relative to $\alpha$-Fe foil at room temperature;

$^b$ Average value in case of distribution;

$^c$ The linewidth used in the distributions varied in the range 0.32–0.55 mm/s;

to them. The fineness of the $\gamma$-$Fe_2O_3$ aggregates could hinder an eventual particular diffractometric pattern. Interestingly, the presence of maghemite, determined by Mössbauer spectroscopy, is restricted to the samples with smaller x values (0 to 0.15) and higher values of MCD.

For samples with $x \geq 0.30$ and MCD $\leq 40$ μm, the $\gamma$-$Fe_2O_3$ phase is not present anymore, a fact revealed by the excellent fits obtained using only two magnetic components and a doublet, all of them respective to the magnetite (Fig. 5b). As earlier reported, incorporation of transition metals such as nickel, cooper and cobalt may increase the resistance of the magnetite to oxidation [13]. Pure magnetite is unstable in aerial environment and gradual oxidation of $Fe^{2+}$, octahedral B-site, is observed within days or weeks after synthesis [10].

Since the magnetic hyperfine field of the site B is progressively being reduced, the doublet is also ascribed to the iron in octahedral positions, where local fluctuations in the lattice (chemical or structural) account for magnetic disorder. Actually, the hyperfine magnetic fields of both sites decrease ($B_{hf}^A$ slower than $B_{hf}^B$), indicating the progressive weakening of the (super)exchange interaction intra and inter magnetic sublattices. For $x \approx$ 0.90, the hyperfine fields are virtually collapsed: the spectrum of the richest zinc sample shows a strong paramagnetic component, which responds for more than 47% of the total spectral area (Fig. 5c) [8, 9]. The same behavior had already been observed for cooper, cobalt and chromium magnetites [11], and reveals how the $B_{hf}^B$ is affected increasing the number of $Zn^{2+}$ ions substituting $Fe^{3+}$ ions as nearest neighbors at site A. The doublet is

similar to the franklinite pattern, for which the ferric and zinc cations completely occupy the site B and the site A, respectively. It is worthy of note that the magnetic distribution of this spectrum may superimpose components of the sites A and B, since the collapse of the first magnetic pattern is inevitable at this concentration.

The evolution of the subspectral areas with the zinc content is shown in Fig. 6. The continuous line represents the theoretical relative area of the site B, with the assumption that zinc enters only at site A and that magnetite is a single phase [11, 12]. The symbols are experimental fitted values. It can be seen that for small zinc concentrations both subspectral areas, for sites A and B, are under the expected values [i.e., $(1-x)/(3-x)$ and $2/(3-x)$, respectively].

This is because the presence of maghemite, which constitute a reasonable fraction of the spectra until to disappear for $x \geq 0.30$. However, for every zinc concentration, the nominal ratio of areas, $2/(1-x)$, between the sites A and B, is reasonably followed.


## 4 Conclusions

$Fe_{3-x}Zn_xO_4$ magnetites were synthesized by co-precipitation. For the lower $x$ values, maghemite was also formed as a minor phase, due to oxidation of the very fine as-prepared magnetite particles. However, as the zinc content increases, only the magnetite phase is observed, showing that substitution favors the chemical stabilization of the $Fe_{3-x}Zn_xO_4$ solid solutions. These solid solutions presented smaller cell parameters, $a$, mean crystalline diameter, $d$, and mass specific magnetic susceptibility, $\chi_{1f}$, with increasing iron for zinc substitution.

All through substitution, the subspectral areas reflected with good agreement the theoretical occupation formula $\left(Zn_x^{2+}Fe_{1-x}^{3+}\right)\left[Fe_{1-x}^{2+}Fe_{1+x}^{3+}\right]O_4$.

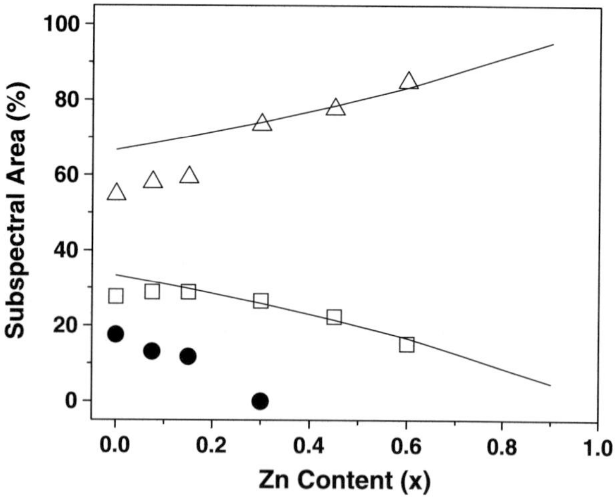


**Fig. 6** Mössbauer relative areas, as a function of zinc content, for the maghemite (*circle*) and magnetite iron sites (*square* = type A; *triangle* = type B). *Continuous lines* represent the theoretical fractions for the magnetites sites, in the absence of any other phase

## References


1. Campbell, S.J., Kaczmarek, W.A., Hoffmann, M.: Hyperfine Interact. **126**, 175 (2004)
2. Benjamin, M., Hayes, K.F., Leckie, J.O.: J. Water Pollut. Control. Fed. **54**, 1472 (1982)
3. Cornell, R.M., Schwertmann, U.: The iron oxides: Structure, properties, reactions, occurrence and uses. Verlag Chemie, Weinheim (1996)
4. Murad, E., Johnston, J.H.: Iron oxides and oxyhydroxides. In: Long, G.J. (ed.) Mössbauer spectroscopy applied to inorganic chemistry, p. 507. Plenum Publ. Corp., New York (1987)
5. Yu, L., Cao, S., Wang, J., Jing, C., Zhang, J.: J. Magn. Magn. Mater. **301**, 100 (2006)
6. Guaita, F.J., Beltrán, H., Cordoncillo, E., Carda, J.B., Escribano, P.: J. Eur. Cer. Soc. **19**, 363 (1999)
7. Li, F.S., Wang, L., Wang, J.B., Zhou, Q.G., Zhou, X.Z., Kumkel, H.P., Williams, G.: J. Magn. Magn. Mat. **268**, 332 (2004)
8. Amer, M.A., El Hiti, M.: J. Magn. Magn. Mat. **234**, 118 (2001)
9. Torres, F., Amigó, R., Asenjo, J., Krotenko, E., Tejada, J.: Chem. Mater. **12**, 3060 (2000)
10. Murad, E., Schwertmann, U.: Clays Clay Miner. **41**, 111 (1993)
11. Sorescu, M., Ihaila-Tarabasanu, D., Diamandescu, L.: App. Phys. Lett. **27**, 2047 (1998)
12. Pereira, S.L., Pfannes, H.D., Mendes Filho, A.A, Miranda Pinto, L.C.B. de and Chíncaro, M.A., Mat. Res. **2**, 231 (1999)
13. Ko, T., Hyun, S., Yoon, H., Han, K., Oh, J.: IEEE Trans. Magn. **41**, 3484 (2005)
14. Wen, M., Li, Q., Li, Y.: J. Electron. Spectrosc. Relat. Phenom. N153, 65
15. Schwertmann, U., Cornell, R.M.: Iron oxides in the laboratory. Preparation and characterization. Verlag Chemie, New York (1991)
16. Dearing, J.: Environmental Magnetic Susceptibility. Using the Bartington MS2 System. Chi Publ., Kenilworth (1994)
17. Wolska, E., Wolniewicz, A.: Phys. Stat. Solid **104**, 569 (1987)
18. Costa, A.C.S.da, Bigham, J.M., Rhoton, F.E., Traina, S.J.: Clays Clays Min. **47**, 466 (1999)

Hyperfine Interact (2007) 175:113–120
DOI 10.1007/s10751-008-9596-x

# Magnetic investigation of iron-nitride-based magnetic fluid

C. B. Teixeira · L. S. F. Olavo · K. Skeff Neto ·
P. C. Morais

Published online: 18 March 2008
© Springer Science + Business Media B.V. 2008

**Abstract** Transmission electron microscopy (TEM), Mössbauer spectroscopy and magnetization measurements were used in the present study to investigate a non-aqueous iron-nitride-based magnetic fluid (MF) sample containing about $2 \times 10^{16}$ particle/cm$^3$. The TEM micrographs indicated spherical-shaped iron nitride nanoparticles with an average diameter of 12.3 nm and diameter dispersity of 0.14. The 77 K Mössbauer spectrum of the frozen MF sample indicated the presence of about 95% of the $\gamma'$-Fe$_4$N phase, with a residual 5% of the $\varepsilon$-Fe$_4$N phase. The temperature dependence of the magnetization was investigated under zero-field-cooled (ZFC) condition, in the temperature range of 4 to 165 K. The ZFC-data were curve-fitted using a new approach that takes into account the particle size distribution and the temperature dependence of the magnetocrystalline anisotropy, making the description of the experimental situation more realistic.

**Keywords** Transmission electron microscopy · Mössbauer spectroscopy ·
Zero-field-cooled magnetization · Magnetic fluids

## 1 Introduction

Magnetic fluids (MFs) are ultra-stable magnetic colloidal suspensions traditionally based on metal-oxide nanoparticles, as for instance, cubic [1] and hexagonal ferrites [2]. The iron-nitride magnetic fluid, on the other hand, has been introduced as a very promising material system for fundamental studies, for it combines high saturation magnetization with enhanced particle size polydispersity characteristics [3]. Iron-nitride MF samples have been considered as an excellent material system to investigate, for instance the "liquid-gas" coexistence as well as the existence of a "ferromagnetic liquid phase" [4]. Preparation of iron-nitride nanoparticles, however, can lead to several magnetic phases, among them the more likely are $\gamma'$-Fe$_4$N and $\varepsilon$-Fe$_4$N [5]. Whereas at room temperature the ultrafine single domain perovskite phase ($\gamma'$-Fe$_4$N) exhibits magnetic ordering the hexagonal phase ($\varepsilon$-Fe$_4$N) shows superparamagnetic behavior [5]. In this study a petroleum-based magnetic fluid sample, containing nanosized

C. B. Teixeira · L. S. F. Olavo · K. Skeff Neto · P. C. Morais (✉)
Núcleo de Física Aplicada, Instituto de Física, Universidade de Brasília, Brasília, DF 70910-900, Brazil
e-mail: pcmor@unb.br

**Fig. 1** TEM micrograph of the iron-nitride nanoparticles. The horizontal bar is 50 nm long

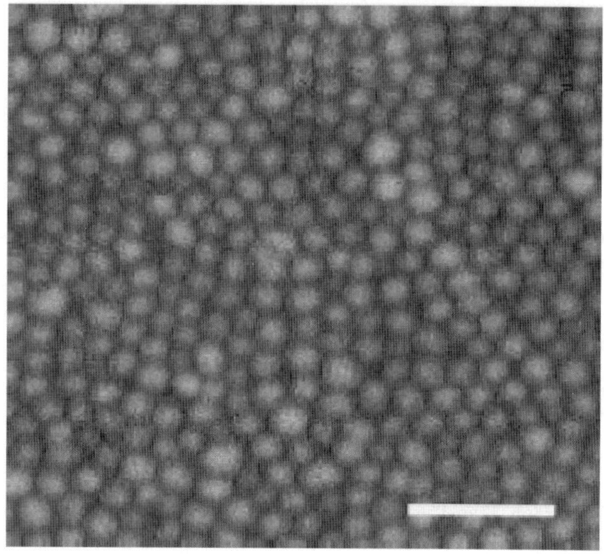

iron-nitride particles, was investigated using the temperature dependence of the magnetization. In order to support the present investigation the magnetic fluid sample was characterized using transmission electron microscopy (TEM) and Mössbauer spectroscopy.

## 2 Experimental

The petroleum-based iron-nitride MF sample was successfully prepared by modifying the synthesis route described in [3]. In short, the nanoparticle synthesis was performed in a dry nitrogen cleansed 500 cm$^3$ flask with the reaction medium stirring at 400 min$^{-1}$. The 1/2 l flask was filled with 60 g of petroleum (195°C b.p.) and 10 g of polyisobutylene succinimid (1,300 m.w.). Ammonia gas was rinsed through the reaction medium at 400 cm$^3$/min while freshly distilled iron pentacarbonyl was added drop-by-drop. The reaction temperature was increased to 90°C and held there for about 1 h. During this first-hour running time non-magnetic intermediate nanoparticles were produced. The reaction medium was then heated to and maintained at 180°C for one extra hour, after which a stable black magnetic colloidal suspension containing about 10$^{17}$ particle/cm$^3$ was obtained. The stock MF sample was then diluted to produce a sample containing $2 \times 10^{16}$ particle/cm$^3$.

The synthesized iron-nitride material revealed spherical-shaped nanoparticles, nearly monodisperse in size distribution, as shown by the transmission electron microscopy (TEM) micrograph (see Fig. 1). The curve-fitting of the particle size histogram (see Fig. 2), using a log-normal distribution function, was accomplished with a mean diameter of about 12.3 nm and size-dispersity of 0.14. Figure 3 shows the Mössbauer spectrum of the MF sample frozen at 77 K. The Mössbauer spectrum was recorded in the transmission geometry, using a MCA (256 channels) and a Wissel constant acceleration transducer coupled to a 50 mCi $^{57}$Co/Rh source. The spectrum in Fig. 3 was curve-fitted using four sextets: three sextets corresponding to 95% of the sample and identified with the $\gamma'$-Fe$_4$N phase and one sextet corresponding to 5% of the sample content and identified with the $\varepsilon$-Fe$_4$N phase [5].

 Springer

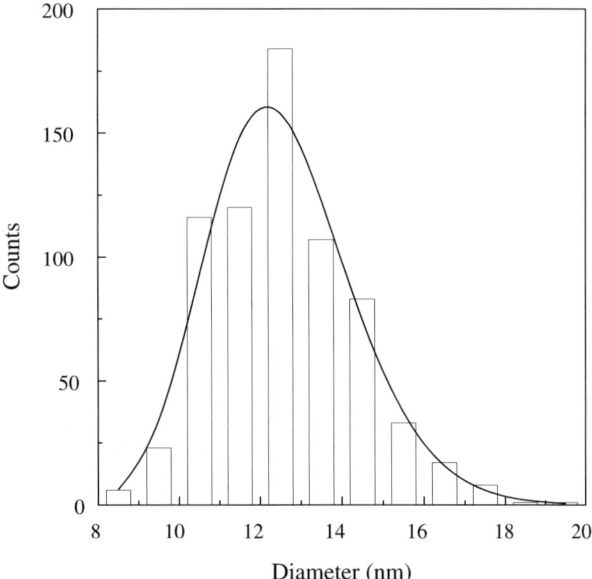

**Fig. 2** Particle diameter distribution of the iron-nitride nanoparticles. The histogram represents the experimental values whereas the solid line is the best curve-fitting using the log-normal distribution function

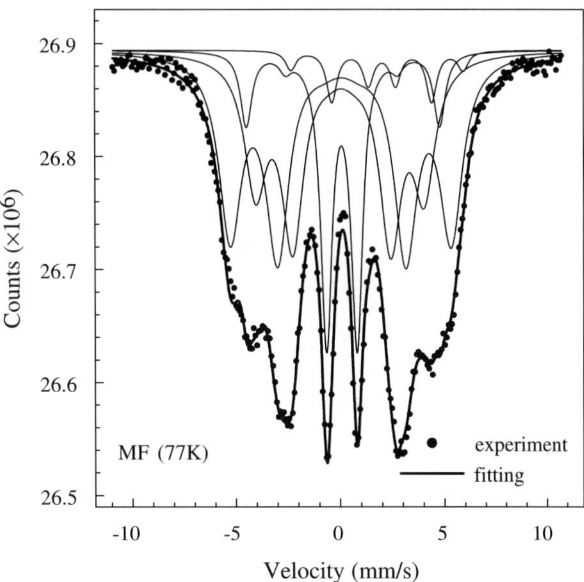



**Fig. 3** Mössbauer spectrum of the frozen iron-nitride-based magnetic fluid sample at 77 K

Solid symbols in Fig. 4a and b are experimental points obtained from ZFC measurements recorded under 300 and 2,000 Oe, respectively. In order to perform the ZFC measurements the sample was first cooled down to low temperatures in the absence of any magnetic field. Then, at the lowest temperature, a DC field was applied to the sample and the magnetization was measured as the temperature of the sample was increased.

 Springer

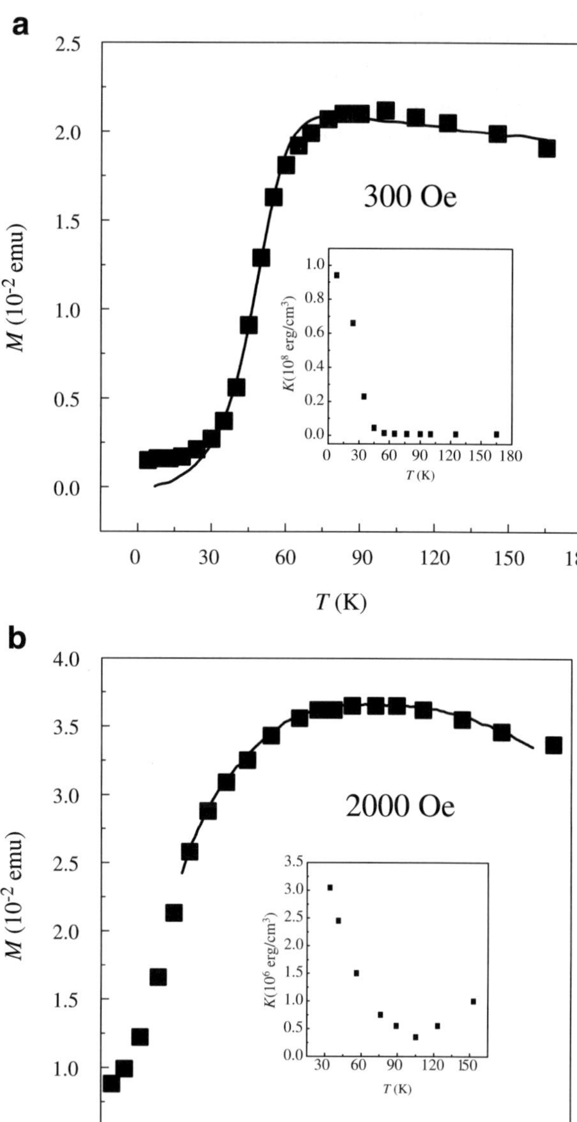

**Fig. 4** Symbols **a** and **b** are experimental points obtained from ZFC magnetization measurements recorded under 300 and 2,000 Oe, respectively. The insets represent the values of the anisotropy obtained from the fitting of the ZFC magnetization data



## 3 Discussion

Analyses of the temperature dependence of ZFC magnetization data, taken from material systems based on nanosized magnetic particles, have been carried out using different approaches. The physical system under consideration consists of a bulk of matter composed of magnetic monodomains described by a magnetic moment $\mu$, which are non-interacting nanoparticles of volume $V$, each one with an uniaxial anisotropy $K(T)$. Under the action of an

 Springer

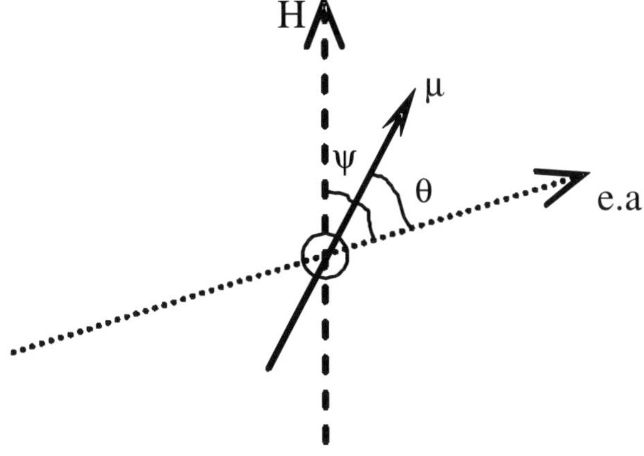

**Fig. 5** Papusoi's coordinate system



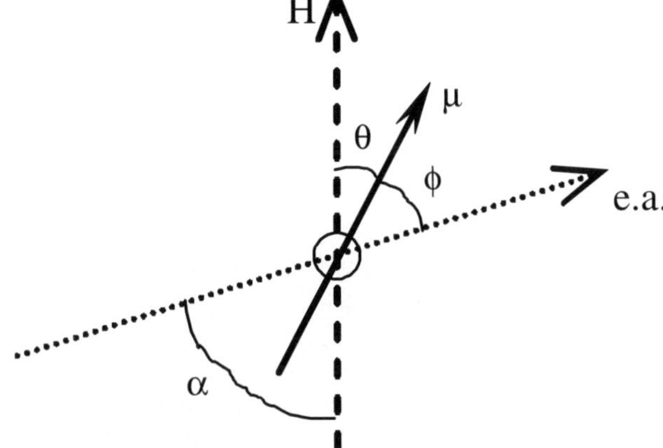

**Fig. 6** Mamiya and Nakatani's coordinate system



external magnetic field $H$ each nanoparticle can be described by a free energy term depending, basically, upon two components: the interaction with the external magnetic field ($\mu H$) and the effective anisotropy ($KV$). One must then present a model picture capable of describing the magnetization curves experimentally obtained, given this energy configuration.

There are some proposals in the literature providing explanation for the ZFC magnetization data regarding superparamagnetic particles. Papusoi et al. [6] assumes the configuration shown in Fig. 5, where $\theta$ is the angle between the magnetic moment of the particle and the easy axis whereas $\psi$ is the angle between the easy axis and the direction of the applied magnetic field. Under this approach the system's energy becomes $E = E_0[\sin 2\theta - 2h\cos(\psi - \theta)]$, with $E_0 = KV$, $h = H/H_a$ and $H_a = 2K/M_{nr}$. $M_{nr}$ is the magnetization of the particle without any relaxation effect, that is considered simply as the saturation magnetization. The Hamiltonian of the problem may be written in terms of $\theta$ and may have two minima with respect to $\theta$, given the values of the other parameters. As a way to identify the two possible situations, the authors define the critical field $h_c = \left[(\sin \psi)^{2/3} + (\cos \psi)^{2/3}\right]^{-3/2}$. If $|h| < h_c$ the free energy displays

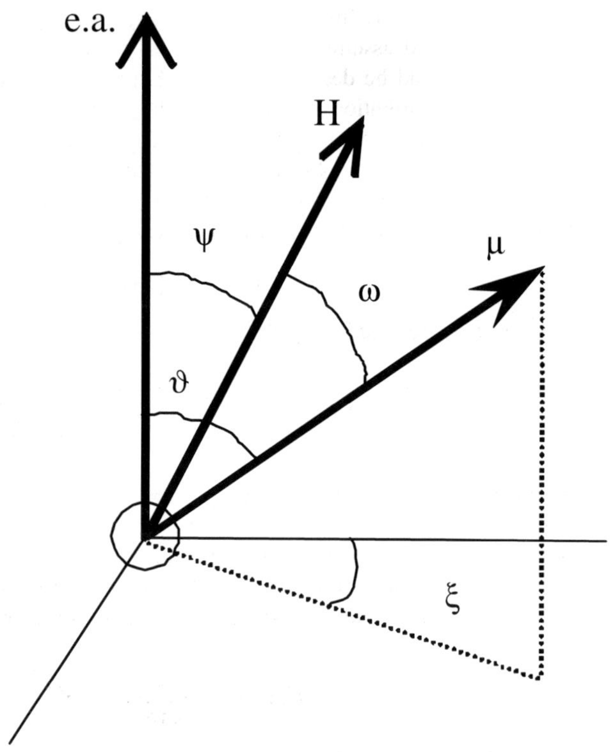



**Fig. 7** Cregg and Bessais' coordinate system

two minima whereas for $|h|>h_\text{c}$ there is only one minima. Nevertheless, for the sample investigated here it is impossible to use the approach reported on [6], once for typical values of $K$ and $M_\text{nr}$ and field values are above 700 Oe one violates the two-minima criterion. In fact, in the very conception of the model it was already foreseen its use at low magnetic fields.

Mamiya and Nakatani [7] proposed a slightly different approach to treat the problem highlighted above, namely the explanation for the ZFC magnetization data of superparamagnetic particles. According to these authors the magnetization is calculated using: $M = \sum_i \mu_i \left[\int \cos\theta_i e^{-E_i/k_BT} d\Omega_i\right] / \left[\int e^{-E_i/k_BT} d\Omega_i\right]$, where $E_i = -\mu_i H\cos\theta_i + KV_i\sin^2\phi_i$. Angles $\theta_i$ and $\phi_i$ are defined in Fig. 6. However, this model includes neither the particle volume distribution nor the orientation of the anisotropy axes in an explicit way. Once a typical sample would have some $10^{15}$ particles, the summation on which the model relies becomes impossible.

Finally, a third approach to the problem of describing the ZFC magnetization data of superparamagnetic-based material systems has been proposed by Cregg and Bessais [8]. In this approach the magnetization of a superparamagnetic system is given by $M(\beta,\alpha,\psi) = M_\text{s}$ $<\cos\omega> = M_\text{s}(\partial Z/\partial\beta)/Z$, with $\alpha = KV/k_BT$, $\beta = HM_\text{s}V/k_BT$, where $K$ is the anisotropy constant, $H$ is the applied magnetic field, $M_\text{s}$ is the saturation magnetization and $V$ is the volume of the particle (see Fig. 7 for variables). The partition function is thus given by $Z = \frac{1}{4\pi} \int_0^{2\pi} \int_0^{\pi} e^{-\alpha \sin^2\vartheta + \beta\cos\omega} \sin\vartheta d\vartheta d\xi$. Difficulties of performing the necessary integrations are well known in this area. In fact, such difficulties are faced in [8, 9], wherein the authors try to simplify the integration problem using series expansions in terms of Bessel functions.

 Springer

In the approach we are first introducing in the present study, we started from the Mamyia–Nakatami model and assumed that the volume of the particles and the orientation of the anisotropy axes could be described by probability distribution functions $F_V(V)$ and $F_\alpha(\alpha)$, respectively. The distribution $F_V(V)$ can be obtained from direct observation via transmission electron microscopy (TEM) whereas $F_\alpha(\alpha)$ can be guessed from the physical characteristics of the problem. Then, the difficulty introduced by the summation in the Mamyia–Nakatami model, considered the main obstacle for its use with bulk matter, becomes an integration that can be easily performed. We started with the energy term given by:

$$E = -\mu(V)H\cos\theta + KV\sin^2(\alpha - \theta), \tag{1}$$

with the scheme described in Fig. 6. Therefore, the expression for the magnetization becomes:

$$M = \int \mu(V)F_V(V)F_\alpha(\alpha)\frac{\int e^{\frac{\mu(V)H\cos\theta - KV\sin^2(\alpha-\theta)}{k_BT}}\cos\theta\sin\theta d\theta}{\int e^{\frac{\mu(V)H\cos\theta - KV\sin^2(\alpha-\theta)}{k_BT}}\sin\theta d\theta}dVd\alpha. \tag{2}$$

For homogeneous anisotropy, one has $F_\alpha(\alpha) = 1/2\pi$. Homogeneous anisotropy is the ZFC case, once the sample is under zero DC magnetic fields and thus, the anisotropy axes must be randomly oriented. When the sample is cooled down, this random distribution of orientation axis is kept as describing the sample. For the volume distribution function we used the TEM data and fitted the data with a log-normal distribution function, given by:

$$F_V(V) = \frac{A}{\sqrt{2\pi}\sigma V}e^{\frac{-\ln^2(V/V_C)}{2\sigma^2}}, \tag{3}$$

where $\sigma$ is the standard deviation, $V_C$ is the average volume, and $A$ is a normalizing constant. As a strategy to calculate the integrals in (2) we adopted the Monte Carlo method. With the change of variables $\mu = M_SV$, $V = V_Cx$, $\beta = \frac{M_SHV_C}{k_BT}$ and $\zeta = \frac{K}{M_SH}$ we rewrite (2) as:

$$M = \frac{M_SV_C}{\sqrt{2}\pi^{3/2}\sigma}\int\limits_0^{V_m/V_C}\int\limits_0^\pi e^{\frac{-\ln^2(x)}{2\sigma^2}}\frac{\int e^{\beta x\cos\theta - \beta x\zeta\sin^2(\alpha-\theta)}\cos\theta\sin\theta d\theta}{\int e^{\beta x\cos\theta - \beta x\zeta\sin^2(\alpha-\theta)}\sin\theta d\theta}dxd\alpha, \tag{4}$$

which is a two-dimensional integral having one-dimensional integrals as the integrand. The Monte Carlo program is quite effective for these two dimensional integrations and the one-dimensional integrals can be quickly performed by usual integration methods. Thus, to calculate the value of the magnetization moment for some fixed temperature $T$ and external field $H$, one uses a Monte Carlo program to choose acceptable values of $x$ and $\alpha$ in (4). With these values the parameters $\zeta$ and $\beta$ are fixed and the one-dimensional integrals can be performed using some traditional method (we used an integration method of order $q^7$, where $q$ is the size of the sub-intervals). Then, the Monte Carlo program chooses another acceptable values of $x$ and $\alpha$ and all the calculations are repeated. Usually, the number of points needed by the Monte Carlo program will be rather small and the computational effort minimized. In fact, for a typical case, magnetization curves for a fixed value of $H$ can be obtained in only a few minutes. Another interesting aspect of the Monte Carlo calculation is that it can provide the error bars of the computational process. We have used a number of points in the integration that gives error bars less than $10^{-3}$.

Solid lines in Fig. 4a and b represent the fittings of the ZFC magnetization data (solid symbols) for 300 and 2,000 Oe, respectively. Solid symbols in the insets of Fig. 4a and b represent the anisotropy values obtained from the data fitting procedure described by (4).

The temperature dependence of the anisotropy as described in insets of Fig. 4a and b is typical of many magnetic materials. The presence of a minimum in the anisotropy versus temperature curve, as clearly observed in the inset of Fig. 4b, is well known in ferrites [10].

## 4 Conclusion

The iron-nitride-based magnetic fluid investigated in this study was morphologically characterized by transmission electron microscopy whereas the magnetic phase of the sample was mainly (95%) composed by the $\gamma'$-Fe$_4$N phase, as indicated by Mössbauer spectroscopy. The $\gamma'$-Fe$_4$N-based nanoparticles were then investigated by zero-field-cooled (ZFC) magnetization measurements in the range of 4 to 165 K. The model proposed in this study to fit the ZFC magnetization measurements may present advantages with respect to other methods in the literature, since it uses a more complete function to represent the sample magnetization, being still computationally fast. In fact, as one of its outcomes, one also ends up with a description of the temperature dependence of the anisotropy. However, the calculation for lower values of the temperature is still a problem.

**Acknowledgment** We thank Dr. N. Buske for the preparation of the iron-nitride MF sample. This work was supported by the Brazilian agency CNPq/MCT and FINATEC.

## References

1. Morais, P.C., Garg, V.K., Oliveira, A.C., Silva, L.P., Azevedo, R.B., Silva, A.M.L., Lima, E.C.D.: J. Magn. Magn. Mater. **225**, 37 (2001)
2. Muller, R., Hiergeist, R., Steinmetz, H., Ayoub, N., Fujisaki, M., Schuppel, W.: J. Magn. Magn. Mater. **201**, 34 (1999)
3. Nakatani, I., Hijikata, M., Ozawa, K.: J. Magn. Magn. Mater. **122**, 10 (1993)
4. Mamiya, H., Nakatani, I., Furubayashi, T.: Phys. Rev. Lett. **84**, 6106 (2000)
5. Panda, R.N., Gajbhiye, N.S.: IEEE Trans. Magn. **34**, 542 (1998)
6. Papusoi Jr., C., Stancu, A.L., Dormann, J.L.: J. Magn. Magn. Mater. **174**, 236 (1997)
7. Mamiya, H., Nakatani, I.: IEEE Trans. Magn. **34**, 1126 (1998)
8. Cregg, P.J., Bessais, L.: J. Magn. Magn. Mater. **202**, 554 (1999)
9. Cregg, P.J., Bessais, L.: J. Magn. Magn. Mater. **203**, 265 (1999)
10. von Aulock, W.H.: Handbook of Microwave Ferrite Materials, p. 280, 366, 503. Academic, New York, New York (1965)

# Author Index to Volume 175 (2007)